MANAGING GLOBAL GENETIC RESOURCES

Forest Trees

MANAGING GLOBAL GENETIC RESOURCES

Forest Trees

Committee on Managing Global Genetic Resources:
Agricultural Imperatives

Board on Agriculture
National Research Council

NATIONAL ACADEMY PRESS
Washington, D.C.   1991

NATIONAL ACADEMY PRESS • 2101 Constitution Avenue, NW • Washington, DC 20418

NOTICE: The project that is the subject of this report was approved by the Governing Board of the National Research Council, whose members are drawn from the councils of the National Academy of Sciences, the National Academy of Engineering, and the Institute of Medicine. The members of the committee responsible for the report were chosen for their special competences and with regard for appropriate balance.

This report has been reviewed by a group other than the authors according to procedures approved by a Report Review Committee consisting of members of the National Academy of Sciences, the National Academy of Engineering, and the Institute of Medicine.

This material is based on work supported by the U.S. Department of Agriculture, Agricultural Research Service, under Agreement No. 59-32U4-6-75. Additional funding was provided by Calgene, Inc.; Educational Foundation of America; Kellogg Endowment Fund of the National Academy of Sciences and the Institute of Medicine; Monsanto Company; Pioneer Hi-Bred International, Inc.; Rockefeller Foundation; U.S. Agency for International Development; U.S. Forest Service; W. K. Kellogg Foundation; World Bank; and Basic Science Fund of the National Academy of Sciences, contributors to which include the Atlantic Richfield Foundation, AT&T Bell Laboratories, BP America, Inc., Dow Chemical Company, E.I. duPont de Nemours & Company, IBM Corporation, Merck & Co., Inc., Monsanto Company, and Shell Oil Company Foundation.

Library of Congress Cataloging-in-Publication Data

Forest trees / Committee on Managing Global Genetic Resources:
     Agricultural Imperatives, Subcommittee on Managing Plant Genetic
     Resources, Forest Genetic Resources Work Group : Board on
     Agriculture, National Research Council.
          p.   cm.—(Managing global genetic resources)
     Includes bibiliographical references and index.
     ISBN 0-309-04034-5 : $19.95
     1. Forest genetic resources conservation.   2. Trees—Germplasm
     resoures.   I. National Research Council (U.S.), Forest Genetic
     Resources Work Group.   II. Series.
     SD399.7.F67   1991
     634.9—dc20
                                                         91-8979
                                                           CIP

Any opinions, findings, conclusions, or recommendations expressed in this publication are those of the author(s) and do not necessarily reflect the view of the organizations or agencies that provided support for this project.

Printed in the United States of America

Committee on Managing Global Genetic Resources: Agricultural Imperatives

PETER R. DAY, *Chairman*, Rutgers University
ROBERT W. ALLARD, University of California, Davis
PAULO DE T. ALVIM, Comissão Executiva do Plano da Lavoura Cacaueira, Brasil*
JOHN H. BARTON, Stanford University
FREDERICK H. BUTTEL, Cornell University
TE-TZU CHANG, International Rice Research Institute, The Philippines
ROBERT E. EVENSON, Yale University
HENRY A. FITZHUGH, International Livestock Center for Africa, Ethiopia†
MAJOR M. GOODMAN, North Carolina State University
JAAP J. HARDON, Center for Genetic Resources, The Netherlands
DONALD R. MARSHALL, Waite Agricultural Research Institute, Australia
SETIJATI SASTRAPRADJA, National Center for Biotechnology, Indonesia
CHARLES SMITH, University of Guelph, Canada
JOHN A. SPENCE, University of the West Indies, Trinidad and Tobago

Genetic Resources Staff

MICHAEL S. STRAUSS, *Project Director*
JOHN A. PINO, *Project Director‡*
STEVEN KING, *Research Associate*
JOSEPH J. GAGNIER, *Senior Project Assistant*

---

* Executive Commission of the Program for Strengthening Cacao Production, Brazil.
† Winrock International, through January 1990.
‡ Through June 1990.
§ Through June 1989.

Subcommittee on Plant Genetic Resources

Board on Agriculture

The National Academy of Sciences is a private, nonprofit, self-perpetuating society of distinguished scholars engaged in scientific and engineering research, dedicated to the furtherance of science and technology and to their use for the general welfare. Upon the authority of the charter granted to it by the Congress in 1863, the Academy has a mandate that requires it to advise the federal government on scientific and technical matters. Dr. Frank Press is president of the National Academy of Sciences.

The National Academy of Engineering was established in 1964, under the charter of the National Academy of Sciences, as a parallel organization of outstanding engineers. It is autonomous in its administration and in the selection of its members, sharing with the National Academy of Sciences the responsibility for advising the federal government. The National Academy of Engineering also sponsors engineering programs aimed at meeting national needs, encourages education and research, and recognizes the superior achievements of engineers. Dr. Robert M. White is president of the National Academy of Engineering.

The Institute of Medicine was established in 1970 by the National Academy of Sciences to secure the services of eminent members of appropriate professions in the examination of policy matters pertaining to the health of the public. The Institute acts under the responsibility given to the National Academy of Sciences by its congressional charter to be an adviser to the federal government and, upon its own initiative, to identify issues of medical care, research, and education. Dr. Samuel O. Thier is president of the Institute of Medicine.

The National Research Council was organized by the National Academy of Sciences in 1916 to associate the broad community of science and technology with the Academy's purposes of furthering knowledge and advising the federal government. Functioning in accordance with general policies determined by the Academy, the Council has become the principal operating agency of both the National Academy of Sciences and the National Academy of Engineering in providing services to the government, the public, and the scientific and engineering communities. The Council is administered jointly by both Academies and the Institute of Medicine. Dr. Frank Press and Dr. Robert M. White are chairman and vice-chairman, respectively, of the National Research Council.

Preface

Forest trees are integral parts of human society. They provide fuel, fiber, construction and building materials, food, and medicines, among other things. The forest itself is an ecosystem, and as ecosystems, forests stabilize environments and are essential components of the global ecology. Although trees are the dominant vegetation, forests are rich reservoirs of biological diversity. They harbor a major proportion of the world's animal and plant species.

Forest trees also enhance and protect our landscapes. They sustain wildlife, industry, and rural economies, and contribute to the quality and richness of our environment.

For many years, concern has been expressed about rapid and continuing losses of the world's forests. In temperate regions of Europe and North America, the decline in forest health has been attributed to industrial consequences, such as acid rain. In the humid tropics, increased demands on the land resulting from the clearing of forests to accommodate expanding populations and the production of industrial products have been highlighted. The causes notwithstanding, the future for the world's forests, if unprotected, is dim.

For many forested areas, efforts to halt or slow losses through the establishment of protected areas will be essential. However, more than protection will be needed. Society will continue to need the services and products derived from the forest. As natural stands of trees are lost, greater efforts to conserve those remaining will continue. There will be more need to select and develop trees and forests that are managed for production purposes to reduce pressure on the remaining natural forests and to provide raw materials in forms more suitable for commercial use and the needs of human society.

To develop these future forests, the genetic resources of forest trees must be conserved and developed, whether they exist as trees in planted or protected conservation stands or as seeds or tissue cultures in storage (or one day possibly as DNA libraries). Managing forest genetic resources involves developing overall strategies, applying specific methodologies, developing new techniques, and coordinating local, national, regional, and global efforts.

Although tree species are similar in many ways to crop species, managing forest genetic resources is not simply a matter of applying programs for crops to larger plants. Forests generate a wide variety of product values, from the different components of the ecosystems that depend on their structural viability to the industrial and agroforestry crops that can be consumed, and the variety of systems that are used to manage them dictates the variety of ways that the genetic resources are used. Moreover, the genetic architecture of forest trees is poorly known, breeding is slow, and wide variations in ecological and economic environments must be anticipated to use the available genetic variation efficiently.

This report, while recognizing serious threats to all the species and ecosystems represented in the world's forests, focuses on managing those forest trees from which harvested materials are currently extracted. Protection of the world's forest ecosystems will require broad efforts by scientists and policymakers. A report prepared by the Commission on Life Sciences in cooperation with the Board on Agriculture, entitled *Forestry Research: A Mandate for Change* (National Research Council, 1990), recommends modifications in the way forestry research is conducted to bring about many needed improvements in the forestry research community. Careful, well-coordinated management of those tree species of current or future harvest potential could play an important part in reducing deforestation pressures on the remaining pristine vegetation. Although the total number of known forest tree species exceeds 50,000 and potential extractive use can be made of several thousand species, current efforts to manage trees focus primarily on fewer than 140 species. Clearly, greater efforts and coordination are needed to sustain even the current levels of forest production and to realize the potential productivity of as yet undeveloped species.

The Committee on Managing Global Genetic Resources, established by the National Research Council in November 1986, is concerned with genetic resources of identified economic value. These resources are important to agriculture, forestry, fisheries, and industry. The committee has been assisted by two subcommittees and several work groups that gathered information or prepared specific reports. One of the work groups, chaired by Gene Namkoong, examined the management of forest genetic resources and drafted this report. It is one of five reports to be published in a series entitled, *Managing Global Genetic Resources.* The other reports prepared by the committee, its subcommittees, and work groups address issues related to the management of plant genetic resources by the U.S. National Plant Germplasm System and the global management of livestock, fish and shellfish, and crop plants. The examination of crop plants will be included in the committee's main report, which will address the legal, political, economic, and social issues surrounding global genetic resources management as they relate to agricultural imperatives.

In addition, a work group was appointed to provide information that would aid in planning and designing a new storage facility for the U.S. National Seed Storage Laboratory. The committee released the report, *Expansion of the U.S. National Seed Storage Laboratory: Program and Design Considerations*, in April 1988. Copies of this report are available from the Board on Agriculture.

The Forest Genetic Resources Work Group was asked to do the following:

• Examine the uses and status of forest tree genetic resources globally.
• Examine in situ and ex situ methods for the conservation of forest genetic resources.
• Assess current germplasm conservation activity for forest tree genetic resources by national, regional, and international organizations.
• Identify the major problems in implementing action, including coordination, information dissemination, and training.
• Recommend future actions to solve or alleviate technical and financial problems.
• Present a global strategy for conserving and managing forest tree genetic resources.

Chapter 1 of this report presents data on the current status of the world's forests and highlights the critical need for conserving and managing the remaining diversity of tree genetic resources, especially in tropical areas of the world. Chapter 2 describes the benefits derived from trees globally and identifies the current and future importance of tree genetic variation in land-use systems and breeding programs. Chapter 3 profiles the current understanding of the biological factors that determine the structure of genetic variation of managed and unmanaged forest tree populations and points out how this knowledge is important to efforts to create, monitor, conserve, and manage reserves for tree species. Chapter 4 defines the methods and technology available for the management of trees through complementary in situ and ex situ conservation activities, and outlines the importance of greater emphasis on developing long-term in situ conservation programs. Chapter 5 describes the activities of the national, regional, and international organizations that are involved with the management and conservation of tree genetic resources. Chapter 6 presents the committee's recommendation on how to implement rapidly a much needed global strategy for conserving and managing forest tree genetic resources.

The sciences related to genetics, plant breeding, and resource management are advancing rapidly. The commitment of increasing numbers of nations to genetic resources conservation is growing. The committee believes that the conclusions and recommendations presented in this report can contribute to improved conservation and management of a major resource—the world's forest trees.

PETER R. DAY, *Chairman*
Committee on Managing
Global Genetic Resources:
Agricultural Imperatives

Acknowledgments

Many people and agencies have contributed their support, time, and creative analysis to this report. Because of the global nature of the study, it was critical that scientists from numerous countries took the time to respond to a variety of questions posed by the committee. To all of those researchers, special thanks are offered.

The committee would like to thank Robert Kellison, Frank Santamour, Kari Keipi, John Spears, and Carl Gallegos for providing initial advice and direction for this report. In particular, the help of John C. Gordon, Raymond Guries, Colin Hughes, R. Sniezko, Milton Kaneshiro, Lilia Barrientos, and Lucy Nunnally is gratefully acknowledged. Special thanks are offered to Christel Palmberg of the Food and Agriculture Organization of the United Nations and to Stanley Krugman of the U.S. Forest Service, who contributed their knowledge and time from the beginning of this study. In addition the committee acknowledges Grace Jones Robbins, Nicole L. Kelsey, Philomina Mammen, Sherry Showell, Carole Spalding, and Maryann Tully of the Board on Agriculture staff for their assistance throughout various drafts of the report.

Contents

xiii

Forest Trees

Executive Summary

The world's forests are declining at unprecedented rates. Losses are resulting directly from clearing to open land for agriculture, roads, and settlements, logging for timber, and cutting for fuel. Indirectly, forests are succumbing to the effects of environmental pollution, and they may be further threatened by climatic changes. Together, these causes have been responsible for decimating many of the world's forests, and they threaten to degrade significantly those that remain. Moreover, the burning of trees, shrubs, and other vegetation during land clearing and after logging further contributes to environmental deterioration.

This report addresses the status of one of the forest's most visible resources: the trees. Many of the world's ecosystems depend on trees for vital functions, such as sustaining soil structure and fertility and preventing soil erosion and floods. Trees provide human society with sources for industrial products, construction materials, food, and fuelwood. Perhaps the most important functions performed by trees and forests are ecological. Their absorption of carbon dioxide and release of oxygen through photosynthesis contribute to controlling the levels of greenhouse gases. This process, in turn, helps moderate fluctuations in global temperatures and provides the atmospheric elements essential for all living things.

Genetic diversity is the mainstay of biological stability—it enables species to adapt to changing environments and to survive. No single organism possesses more than a fraction of the genetic variance of the species. The sum of gene differences among scattered populations of a

A rain swollen stream runs through a lush rain forest in the Xishuangbanna region of Yunnan Province, People's Republic of China. Credit: Jodi Cobb ©National Geographic Society.

given species constitutes the gene pool. Thus as the numbers of individuals and populations that comprise a species are reduced and its gene pool is eroded, the species can be pushed toward extinction. Once populations or species are extinct, the genes they possessed can no longer aid their adaptation to changing environments or be used in developing improved varieties.

The status of the world's tree genetic resources and the technologies and institutions for their management and use are assessed in this report. Particular attention is given to those tree species of current or potential economic value. Although several groups and institutions around the world maintain collections of tree germplasm (see Appendix C), there is urgent need to develop greater capacities to collect, conserve, and use tree genetic resources. Developing selected tree species to meet societal needs can reduce the threats to the world's natural forests by providing alternatives to harvesting them. The report recommends actions for maintaining a species within its natural community (in situ management) and for maintaining planted stands in areas outside their normal range and conserving trees as seed, pollen, or tissue cultures (ex situ management). It also proposes directions for research and

Western hemlock trees are nestled around a lake in the temperate rain forest of Olympic National Park, Washington. Credit: Robert W. Madden ©National Geographic Society.

institutions that could address the problems of tree loss at national, regional, and international levels. It also proposes a global strategy centered on an international institution and suggests ways in which such an institution could be constituted.

THE CRISIS OF WORLD FOREST DECLINE

About 30 percent of the world's ice-free land surface is forest or woodland. The forested areas of the world today comprise between 3.8 billion ha (Council on Environmental Quality and U.S. Department of State, 1980) and 4.5 billion ha, or an area equal to the size of North and South America (World Resources Institute et al., 1988). These areas include the moist tropical forests, the most complex, species-rich ecosystems in the world. Removal of the natural vegetation and overcropping of trees are responsible for an estimated total degradation of 2 billion ha of tropical forests (Wood et al., 1982). It has been commonly estimated that in tropical forests alone an area equal to the size of Guatemala, 11 million ha, is lost annually. Not only are $37.5 billion in wood exports to developed nations threatened, but without immediate action, 2.8 billion people in developing countries will be out of adequate fuelwood supplies before the end of the twentieth century (Food and Agriculture Organization, 1985c).

When forests decline or are removed, trees are not all that is lost. The forest harbors many forms of animal and plant life that depend on its environment for survival. Many of these species, their potential use to society, and their ecological importance have yet to be discovered. Once those species are gone, that knowledge, along with the potential benefits, is also gone. If the current rate of decline in tropical America continues, for example, fully two-thirds of its plant species may be extinct by the end of the next century (Simberloff, 1986). Appendix B lists 492 tree species or provenances reported in the literature to be threatened to some degree. Of those species and provenances, 147 are in danger of extinction and 156 are vulnerable to loss.

Less visible than the loss of species, but equally alarming, is the decline in genetic diversity (variation) within forest species. Genetic diversity is the basis for the natural evolution and adaptation of species to new, changing environments. It is the manipulation of this diversity by farmers and scientists that has produced the highly productive and specialized crops and livestock of modern agriculture. Both the survival of natural forests and the development of selected tree resources for industrial and other societal uses require a broad base of genetic diversity.

BENEFITS, USES, AND DEVELOPMENT OF FOREST TREES

The benefits of forest trees to human society are both direct and indirect. Forest trees directly provide such industrial products as fibers, resins, oils, pulp, and paper; pharmaceuticals; building and other construction materials; fodder; and fuelwood. Annual wood production alone from the world's trees is 1.5 billion m^3. More than half the wood used each year becomes fuel for heating or cooking.

The indirect benefits of maintaining viable populations of trees include ecosystem protection and amenity, or social, values. Forest trees are the dominant and necessary prerequisite vegetation for the functioning of many ecosystems. Through photosynthesis, essentially the only mechanism of energy input into the living world, trees use carbon dioxide in the atmosphere to produce the oxygen necessary to support life. They also contribute to developing and maintaining soil structure and fertility. Forests hold soil against erosion, and their degradation or removal has exacerbated such problems as flash flooding and sedimentation in reservoirs.

Trees are found in a broad spectrum of land-use systems, from

Near Kassala, Sudan, women carry bundles of fuelwood. About 80 percent of the population relies on trees for energy as firewood or charcoal. Credit: Food and Agriculture Organization.

planned and managed systems to naturally occurring stands. The variety of such systems can be grouped into three classes: agroforestry (trees, crops, and animals are raised in combination or sequentially), industrial plantation forestry (growing trees for timber), and natural vegetation management (managing naturally occurring stands for production or conservation). Both agroforestry, as for some *Acacia* in Africa, and industrial production forestry, as for *Cryptomeria* (red cedar) in Japan, involve some degree of genetic resource management, including the direct breeding of trees to meet specific needs. The goal of natural vegetation management is the maintenance of the natural state for one or more species of trees. In all of these systems, knowledge of genetic diversity and its organization within a population or species of trees can aid conservation and use.

At present, no adequate global strategy exists for systematically identifying, sampling, testing, and breeding trees with potential use. The development of improved varieties of tree species for use in industry, agroforestry, and the rehabilitation of degraded lands has been given little attention. Nor is there a unified list of trees with high potential for use. The number of tree species that could be candidates for development would at least double the 400 species identified by the committee as having been tested or included in breeding programs (Appendix A). At least tenfold more tropical tree species should be tested in breeding programs.

An obvious need exists to plan and implement an inventory of the diversity of the world's forests. Current tree use and management efforts are not generally based on knowledge of the inventory of the world's forest tree species. The larger proportion of tree species currently in testing and breeding programs are from temperate regions (see Appendix A). Efforts are needed to add information on tropical and subtropical forest tree species from south and southeast Asia, tropical Africa, and Latin America. Enormous diversity in hard- and softwood species occurs in evergreen humid tropical and deciduous forests (R. S. Paroda, Indian Council of Agricultural Research, New Delhi, personal communication, October 1989).

At this time, few programs conserve the genetic resources of species that have no current production value and that are outside any ecosystem

In Coos Bay, Oregon, a ship loads logs bound for Japan. Credit: James P. Blair ©National Geographic Society.

reserve or protected area. One notable example is the Central America and Mexico Coniferous Resources Cooperative at North Carolina State University which holds unique collections of several potentially useful species. Conservation of species that serve vital functions for ecosystems, commercial production, recreation, or other social values, but that are not now included in protected areas, has also been neglected. *Tabebuia impetiginosa* in Brazil, for example, is useful for its wood as well as other functions, is vulnerable to loss, but is not being conserved in any in situ or ex situ program. A need exists not only to expand current efforts, but also to coordinate the setting of priorities to increase the number and diversity of species conserved.

STRUCTURE OF GENETIC VARIATION IN TREES

Managing the genetic resources of trees requires an understanding of the biological dynamics of the populations in which they exist. In particular, knowledge of the diversity and distribution of genes in a tree population is crucial to genetic management, because genetic information makes it possible to predict the likelihood of gene loss in a population and to develop strategies to prevent such depletion. Understanding genetic structure in a tree species will make it possible to collect the genetic diversity of natural populations efficiently and to ensure their conservation.

The extent and pattern of genetic diversity in forest trees are strongly influenced by their mating systems and the movement of genes between dispersed populations of the same species (gene flow). Mating systems in forest trees are varied and range from mechanisms to ensure outcrossing (pollination by other individuals in the population), such as among *Pinus* (pine) species, to mechanisms for inbreeding (self-pollination), such as among some *Eucalyptus* species. Designing reserves for trees, that is, their sizes, shapes, and distributions, requires an understanding of reproductive biology. However, knowledge in this area remains limited, especially for tropical species.

Gene flow affects the degree to which individual populations of trees are genetically isolated. An understanding of this is important when designing in situ reserves or ex situ conservation stands to ensure that genetic diversity is preserved. Gene flow, and thus gene distribution, is also important when collecting samples from tree populations to obtain a range of genetic diversity for ex situ conservation.

Measuring genetic diversity in trees has typically been done by either provenance testing or electrophoretic analysis of enzymes. The former involves visual observations of the differences between tree samples grown under similar conditions at different sites. Trees, especially exotics, grown in unsuitable provenances may die, have dieback, or display unsuitable form. The latter compares the number and variety of specific classes of enzymes present in the samples being tested.

Both techniques show that forest trees, in general, contain considerable genetic variation. It is not known, however, to what degree the variation revealed by electrophoretic surveys reflects genetic differences in the capacity of individual trees to compete or adapt. Further, electrophoretic variation can be indicative of genetic variation, but results from electrophoretic analyses also can be affected by physiological states or environmental conditions. Thus, the degree to which electrophoretic data can be correlated with the data from provenance testing is ambiguous; sometimes the data are similar, but often, as in the case of *Pseudotsuga* (Douglas fir), they are quite different (El-Kassaby, 1982).

Despite the heterogeneity of forest trees, genes may not be distributed randomly in a tree population and, thus, sampling of the diversity within a population may be complicated. Current data suggest that for wind-pollinated, widely spread species sampling a small number of populations may be sufficient (Hamrick, 1983), but for species like *Liriodendron tulipifera* (tulip tree) and *Quercus velutina* (red oak), which grow in many isolated populations, sampling a larger number may be required to ensure an unbiased sample of genetic diversity.

Geographic, climatic, and biological factors affect the distribution of genes in a species. For the majority of forest trees, however, knowledge of genetic structure is sparse, and knowledge of the degree to which such characteristics as drought tolerance, growth, and disease resistance are heritable is still deficient, although growing. Such information is necessary to develop reliable and effective methods of sampling the total genetic diversity of tree populations.

RECOMMENDATIONS

Important opportunities for confronting the challenges of managing the world's forest tree resources have been identified. New and increased efforts will be needed in developing and applying genetic management techniques. Research and development are required to address fundamental issues and information needs. Institutional efforts, particularly at the national level, must be strengthened. Finally leadership is urgently needed to develop and facilitate a global strategy for conserving and managing the world's tree genetic resources.

Through afforestation measures, hardy eucalyptus trees are used for sand dune fixation in Senegal. This effort is part of local community development activities and land reclamation projects for agriculture. Credit: Food and Agriculture Organization.

Genetic Management Techniques

A combination of in situ and ex situ techniques will be needed for any global, regional, or national efforts to conserve and manage the genetic resources of trees. In situ and ex situ methods are taken as complementary, not opposing, methodologies. In situ conservation provides the opportunity to preserve the broadest range of species, but ex situ collections may be more appropriate when access to specific, well-studied, or threatened populations is desired.

Increase of In Situ and Ex Situ Programs

In situ and ex situ programs to conserve, manage, and use forest tree resources must be significantly expanded to encompass at least a tenfold increase in the species that are included.

Efforts to conserve and manage tree genetic resources do not encompass global needs. Deficiencies exist both in information and in the extent of activities. For many species of recognized potential value, new efforts are needed for ex situ conservation. In tropical and subtropical regions, where species diversity is greatest, many more species should be conserved in situ.

In Situ Management

Long-term, in situ genetic management plans should be developed, especially for tropical and subtropical species.
In situ management of trees typically refers to conservation in undisturbed nature reserves, managed nature reserves, and national parks. It can also include planned and managed plantings of trees within the native environment of the species. This is somewhat different from in situ management for crop plants, wherein such planted stands are generally considered a form of ex situ conservation.

The design of in situ conservation schemes is still primitive. Little is known of the nature of current and future stresses that can affect even local mortality and extinction rates for most tree species. Until more is known about the population sizes and structures necessary to prevent genetic loss, a large degree of genetic redundancy for conserved species will be needed.

Ex Situ Management

Development and application of technologies for the ex situ conservation of pollen, seed, and tissue cultures, as a supplement to in situ maintenance, should be encouraged.
One example of ex situ management is the organization by several working groups of the International Union of Forestry Research Organizations (IUFRO) of a large number of collections and tests in Europe of species that are native to North America. Ex situ methods also include maintenance of seed, pollen, tissue cultures, or other sources of genetic material for propagation. Seed storage of many tree species is both possible and practiced, but the long times to maturity and genetic instability during regeneration can make obtaining a new crop of fresh seed an expensive and uncertain process. For many important species of genera such as *Quercus* (oak), *Shorea* (mahogany), and *Hopea*, seeds are short-lived or die when dried for storage. Ex situ methods of tissue culture and cryogenic storage (i.e., storage in or suspended above liquid nitrogen at temperatures from $-150°C$ to $-196°C$) could enable long-term maintenance of such species, but further effort is needed to apply these to many tree species. In the future, genetic information in the form of DNA (deoxyribonucleic acid) libraries may be maintained for tree species.

Tree collections held as seed are primarily for short- or medium-term storage for use in afforestation (the planting of trees in unforested areas) and reforestation activities. Few programs have long-term objectives, and at present the techniques for regenerating most of these collections are not clearly defined. Methods for the rapid regeneration of stored seed collections are urgently needed.

Ex situ stands or planted forests should be developed that can serve as living seed banks, as test and evaluation stands, or as both.

The use of ex situ stands of trees for conservation purposes can be applied to many more species than are now included. Relatively small areas can be part of a conservation network that ensures the survival and availability of genetic materials and that provides data on performance over a variety of sites. The integration of such conservation networks with breeding programs and in situ programs could provide a vital link between conserving and effectively using the total gene pool of a species. No global programs to foster such linkages exist, although the Food and Agriculture Organization (FAO) of the United Nations, the Nitrogen Fixing Tree Association, and the Oxford Forestry Institute (OFI) have initiated such programs in a few instances, for example, in *Leucaena* and *Acacia* species.

Exploration, Testing, and Breeding

Additional cooperative efforts among nations are needed to develop coordinated programs for exploring, collecting, and evaluating tree genetic resources.

When a species is indigenous to several countries and is useful in others, coordinated efforts can have many benefits, such as reducing the number of collection missions needed but allowing a broad range of environments to be sampled. Many eucalypts, *Gmelina*, teak, and some tropical pines and hardwoods are among the species included in cooperative programs today.

Forest genetic resources programs should conserve species that lack clear present or potential value and those that have known potential value.

No programs exist to conserve the genetic resources of species that lack clear potential production value or that are not adequately included within ecosystem reserve areas. Few groups or institutions promote the management or conservation of many species that might serve vital functions for ecosystem productivity, recreation, or other diffuse values, but remain outside protected areas.

Breeding programs are needed for many species, especially those new to genetic management programs. For many hundreds of species

In the Amazon River Basin, Brazil, the clearing of rain forest for pastureland has led to erosion and loss of productivity, which in turn prompt more clearing for new pastures. Credit: James P. Blair ©National Geographic Society.

of potential value, genetic surveys and sampling for multiple populations are needed. Developing the species of known potential value will require an order-of-magnitude increase in the number of species included in exploration and testing programs.

The number of species with potential value that are included in testing and breeding programs should be at least doubled.

In production forestry, the greatest efforts at management are focused on fewer than 140 of the more than 50,000 estimated total tree species worldwide. Of those 140 species, slightly more than half are included only in seed collection stands, which are essentially large, mass-selection populations. About 60 species are in breeding programs sufficiently intensive that ex situ conservation is included as part of the program. Only a few of the latter species, such as *Pseudotsuga menziesii* (Douglas fir), *Pinus caribaea*, and *P. oocarpa* (West Indian pines), are included in international cooperation programs for genetic management.

The industrialized countries of the temperate and boreal regions have well-established species of widespread commercial use, some populations of which have been widely sampled in parks, test stands, and breeding and production stands. However, many of these species are incompletely sampled for conservation purposes. All of them lack clear programs for using genotypes or populations as introductions or substitutes for their current breeding population(s). The number of tree species with clear potential for future use that could be managed in breeding or prebreeding operations is at least twice the number of species currently used, and their inclusion would quadruple the number of populations used.

Programs are needed to ensure long-term storage (including cold storage) of tree germplasm and to coordinate efforts to maintain managed stands.
Of the available techniques for gene conservation, heavy reliance is placed on managing living stands, both ex situ and in situ. Programs for long-term storage as seed, pollen, or tissues are very limited. As long as interest remains, stand management should continue, but it is vulnerable to lapses in funding and control and to environmental changes.

Global Data Base

A global data base on the status of tree genetic resources should be established and continuously updated. It should include listings of ongoing conservation activities, breeding programs, test stands, and other activities pertinent to conserving trees.
A data base would facilitate identifying species for conservation or for addition to testing and breeding programs. It would foster interaction among regional and national programs, support and encourage training at the international, regional, and local levels, support research and its application to managing forest tree germplasm, and provide a central source for assembling and disseminating data to national and regional programs. It would thus amplify the considerable efforts of the FAO's Panel of Experts on Forest Genetic Resources.

Research and Development

Genetic Variation

To support conservation efforts, study of the patterns of genetic variation in tree populations should be accelerated and expanded in scope, especially in the tropics and subtropics.
Genetic variation has been surveyed in a wide range of species, but the existing data base is largely derived from the study of conifers in

the north temperate zone. Despite the very different reproductive mechanisms of many tropical species, a preliminary analysis reveals that levels of genetic divergence within populations are similar to those observed for temperate tree species. However, further study is needed before this information can be generally applied to conserving tropical tree species.

Inventory

Increased efforts are needed to provide an accurate inventory of forest trees.
Even as new research is initiated, tree populations are being changed and eroded. The scale of change, both in the area affected and the speed of change, has vastly increased. The loss of species and genetic variation in populations due to deforestation may be slowed if critical populations are targeted and saved. Similarly, the effects of regional population or global climatic changes might be counteracted in the future by constructing new populations from conserved populations and collections. To do this, knowledge of what is in these forests is needed.

Genetic Structures

Research is needed to elucidate the distribution and structure of genetic variation—especially for tropical trees—and to support conservation efforts.
Information on reproductive systems, the genetic architecture of populations, and the requisite population sizes of trees is essential for the design of in situ and ex situ conservation stands. Understanding the subtle factors that create or destroy genetic diversity provides the scientific foundation for managing tree genetic variation. Differences in the mating systems of many tree species and the ecological instability of stand boundaries can strongly affect genetic variation among populations. This must be considered when managing a sampled set of genotypes or conservation stands. However, the interactions of species and the structure of populations, especially in tropical species, are poorly understood. Obtaining such knowledge for tropical regions, where tree species occur in low densities as part of complex ecosystems, can be especially difficult.

Climate Change and Pollution

More comprehensive study should be made of the effects of global climatic change and pollution on forest tree species.

In the temperate regions pollution from industrial and transportation activities, combined with other stresses, is causing severe decline in forest vigor. Trees are dying prematurely in most industrialized countries and in some developing countries. Research efforts are in progress to understand the causal agents and biochemical pathways through which soil, water, and trees themselves are affected by pollution. In Germany alone more than 1,000 government-sponsored projects are examining the effects of pollution on forests. The Executive Board of the IUFRO has established the Special Task Force on Pollution to coordinate research and information on this problem, and reviews of the state of knowledge formed a major part of its Eighteenth World Congress in Montreal during 1990.

Institutions

The current status of tree genetic resource conservation activity is inadequate to prevent future losses in genetic diversity. Major international efforts in the use of forest tree genetic resources originated in the late 1960s, with the guidance and support of the FAO and several national institutions, mostly from developed countries. Since then, institutional interest has been stimulated because tree improvement programs have demonstrated that economic benefits can be realized from using genetic resources to improve forest production and because increasing environmental awareness, including recognition of deforestation and species loss, has led to increased recognition of the need to preserve gene resources.

New Programs

The capacity of national and regional institutions to manage tree genetic resources, especially in areas where loss of species and populations is most severe, should be enlarged and strengthened. Forest genetic resource conservation programs should be formalized and included in national plans for forestry, biological diversity, and breeding or other forms of genetic management.
The in situ and ex situ conservation efforts of national governments require focus, coordination, and support. In the nonindustrialized world demand is growing for production forestry, especially for fuelwood and fodder in dry areas, but also for timber and other wood products. There is a need to establish breeding populations for nearly all species that are prime candidates for such uses. Preliminary breeding programs have been initiated with some previously wild species. However, several hundred species of potential value, especially in arid zones and the

This tropical rain forest at 500 m near Serra dos Carajás is part of the world's largest closed rain forest, estimated to be 357 million ha in size, located in Brazil. Credit: Calvin R. Sperling.

tropics, are not yet included in such programs and continue to be at high risk of loss.

The precise, future global needs for forest genetic resources are not possible to predict, but the widest possible genetic base should be maintained to provide the best array of future options. The programs of national and regional institutions to conserve, describe, document, improve, and distribute forest genetic resources are critical activities, and they urgently need to become more active and more effective. Techniques for establishing priorities must be developed. The particular requirements and challenges of countries and regions may vary, but the overall goal is to conserve and maintain a broad range of forest tree genetic resources for current and future use and as a foundation for conserving the world's biological diversity.

Funding and Personnel

Sustained political support and expanded independent funding must be provided for long-term forest conservation operations, for professional and

Previously forested land in Brazil, near Serra Pelada, has been cleared to accommodate a gold mining settlement. Credit: Calvin R. Sperling.

technical staff training, and for stabilization of institutions that address the needs of tree genetic resource conservation and management.

For many nations the funds available to protect forest resources may be insufficient for genetic objectives. Although it is at the national level that needs and priorities must be elucidated and where "on-the-ground" activities must occur, establishing and fostering such efforts will require bilateral or multilateral political and financial support. National programs will remain responsible for the major efforts in tree conservation, particularly where species of current or future use are concerned. Frequently, such efforts can occur through existing national departments or ministries established to address the needs of national forests, but their programs must be extended to include genetic resources. The education and training of professionals and technicians in forest genetic resource conservation must also be expanded to provide sufficient technical and support staff to meet urgent needs that will result from the increased activity.

Regional institutions could assist member countries in developing national and cooperative activities related to forest genetic resource conservation. They could provide funds for programs on forest genetic resources, inventories, seed collection, nurseries, testing, ex situ stands, scientific exchange within and outside the region, training, and research. Regional programs would be uniquely suited to addressing the needs and facilitating activities for national programs that have common interests and priorities with regard to managing tree resources and the species of concern.

Developing a Global Strategy

Addressing the world's needs for tree genetic resources will require a global strategy that focuses on in situ and ex situ conservation, research, and institution building at national, regional, and international levels.

International Leadership

An international institution should be established or designated to provide leadership, coordination, and facilitation for the global management of the world's tree genetic resources at national, regional, and international levels.

In situ and ex situ strategies should compose complementary elements of an integrated program. With limited resources to accomplish this task, coordination among the many national, regional, and international groups and institutions is needed to focus efforts on critical needs, to avoid unnecessary duplication of effort, and to ensure optimal coverage.

The comprehensive management of the tree resources of the world's forests requires leadership to coordinate existing efforts and foster new ones. Institutions, such as the FAO, OFI, Centro Agronómico Tropical de Investigación y Enseñanza, Centre Technique Forestier Tropical, and others discussed in this report, conduct important work at multiple national and international levels. There are, however, omissions both in priorities and methods of conservation. An international institution to coordinate and to facilitate the global conservation and management of forest genetic resources is required. The main function of such an institution would be to maintain an ongoing assessment of the status of forest genetic conservation worldwide and to foster the study, collection, documentation, evaluation, and utilization of tree genetic resources.

Global leadership could be provided through a shared effort of the FAO and a forestry body created within the International Board for Plant Genetic Resources (IBPGR). These efforts could be augmented by expertise from other international, regional, and national agencies. Such global cooperation would enable the initiation of efforts more rapidly than if an entirely new program were established. The FAO historically has played an important role in developing forest programs and disseminating information. The technical and scientific capacities of the IBPGR would enhance FAO's efforts. For both institutions such an arrangement will require some administrative adjustment and additional funding.

Trees are an integral part of the natural world and they are essential to life. Trees have always been indispensable to the functioning of human society. It has become apparent in recent decades, however, that the world risks losing much of this invaluable resource. If society is to conserve the biological diversity encompassed in the world's trees and the forests and habitats they create, then renewed efforts are needed. The demand by a burgeoning population for the products and benefits of trees requires greater efforts to develop varieties adapted to human needs. Similarly, the remaining world forests, and the diversity they contain, must be protected from the increasing pressure to consume them. This is no longer the responsibility of a few nations, but will be achieved only through a global, cooperative effort.

1

World Forests

About 30 percent of the world's ice-free land surface is occupied by forests and woodlands. Permanent pastures for livestock (24 percent) and arable lands under crop production (11 percent) also contain significant numbers of trees and woody shrubs that provide a variety of services and products for humans (Food and Agriculture Organization, 1988). The decline and loss of the forested portion of the globe have been vast and have caused grave concern among scientists and policymakers worldwide.

The forested areas of the world now comprise between 3.8 billion ha (Council on Environmental Quality and U.S. Department of State, 1980) and 4.5 billion ha (Table 1-1). The coniferous forests of North America and Eurasia cover an estimated 1.3 billion ha, other temperate forests about 1.1 billion ha, and tropical forests about 1.7 billion ha.

LOSSES IN FORESTS

Although major losses have also occurred in temperate forests, the greater concern is about losses in the moist tropical forests, which comprise the most complex, species-rich ecosystems in the world (Office of Technology Assessment, 1984). These forests are being destroyed at

Tree species discussed in this report are identified by their scientific names. The use of common names is complicated because many may exist for a single species, they may refer to only a portion of that species, or none may exist for some species. Where clarity could be improved, common names supplement the scientific names.

TABLE 1-1 Distribution of the World's Forestlands (millions of hectares)

Region	Land Area	Closed Forest		Other Wooded Areas[a]		Total Forest and Wooded Lands[b]	
		Area	Percent of Land Area	Area	Percent of Land Area	Area	Percent of Land Area
World[c]	13,077	2,792	21	1,707	13	4,499	34
Temperate zone	6,417	1,590	25	563	9	2,153	34
North America[d]	1,835	459	25	275	15	734	40
Europe	472	145	31	35	7	181	38
Soviet Union	2,227	792	36	138	6	930	42
Other countries[e]	1,883	194	10	115	6	309	16
Tropical zone	4,815	1,202	25	1,144	24	2,346	49
Africa	2,190	217	10	652	30	869	40
Asia and Pacific	945	306	32	104	11	410	43
Latin America	1,680	679	40	388	23	1,067	64

[a] Includes open woodland and wooded areas with forest regrowth following clearing for shifting cultivation within the past 20 years.
[b] Includes closed forest and other wooded areas.
[c] Excludes Antarctica.
[d] Canada and the United States.
[e] Australia, China, Israel, Japan, New Zealand, and South Africa.

SOURCE: Table adapted from *World Resources 1988–89* by The World Resources Institute. Copyright 1988 by The World Resources Institute and the International Institute for Environment and Development in collaboration with the United Nations Environment Program. Reprinted by permission of Basic Books, Inc., a division of Harper-Collins Publishers.

an unprecedented rate. The commonly accepted estimate (Lanly, 1982) is that the net annual loss amounts to 11 million ha, an area the size of Guatemala. Closed forests in the moist tropics, those that have a mostly continuous canopy that collects a high degree of sunlight, are being lost at an annual rate of 7.5 million ha. Open formations in drier areas are declining by 3.8 million ha annually (Grainger, 1987; Wood et al., 1982). In the dry tropics, semiarid vegetation types and savannah woodlands are being turned into wasteland at a rate of about 2 million ha annually.

About 36 percent of the tropical forests that have been degraded by removal of natural vegetation and overcropping could still be rehabilitated with the introduction of tree species. This includes 418 million ha in dry or montane areas in need of reforestation, 137 million ha of tropical rain forest in need of protected regeneration, and 203 million ha of forest fallows in the humid tropics in need of reforestation (Wood et al., 1982). In India alone, there are over 100 million ha of wastelands, an area the size of Egypt, and vigorous attempts are now being made to revegetate those lands with tree species to restore them to productive use (Hegde and Abhyankar, 1986).

If the deforestation rates in the Caribbean and in Central and South America remain constant, the reduction in forest area by the end of the twentieth century will result in the loss of an estimated 15 percent of all currently identified plant species and 2 percent of all of tropical America's plant families (Simberloff, 1986). If the rate remains steady, it is also predicted that 66 percent of the plant species and 14 percent of the plant families may disappear by the end of the twenty-first century (Simberloff, 1986). These estimates are based on the number of species that are known to science at this time. Research and exploration activities may discover many new plant species, which in turn may increase estimates of the number threatened. Even at the current rate of destruction, catastrophic losses in plant biodiversity are likely and the loss of genetic resources could be substantial.

The loss of forest genetic resources is not confined to the tropics. Especially in Europe and North America, atmospheric pollution and wildfires have threatened forests and the genetic resources of a range of species (Scholz et al., 1989). Degraded forestlands and watersheds may be rehabilitated, and denuded hillsides afforested, but when a species becomes extinct, or genetic variation is reduced, the loss is permanent. Even if forests are regenerated by natural or artificial means, they may become less capable of genetically adapting to environmental challenges and to future large-scale stresses.

The loss of valuable plant genetic material could be even more significant in little known species that may have small distributions but

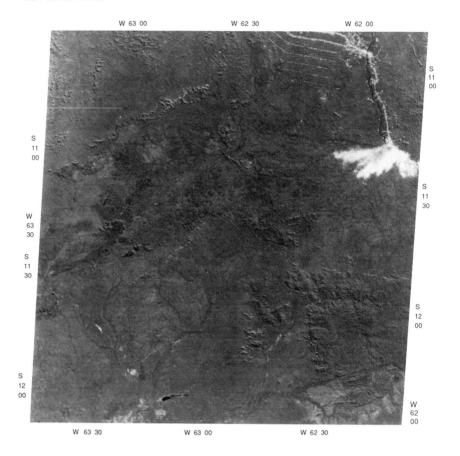

These two satellite images have recorded the effects of the construction of the BR-364 highway in Rondonia, Brazil. The image on the left shows a section of Rondonia on July 7, 1973, with BR-364 appearing as a white track in the upper right. Note the lake in the lower left. Dark areas represent forests. The white area

that may be potential sources of food, drugs, fuel, and fiber (Harlan and Martini, 1936; Landauer, 1945; Myers, 1983a,b; Oldfield, 1984; C. Prescott-Allen and R. Prescott-Allen, 1986; R. Prescott-Allen and C. Prescott-Allen, 1983). Recent estimates have placed about 400 tree species on the list of species that are endangered in whole or in significant parts of their gene pools (Food and Agriculture Organization, 1985a), and the list is growing rapidly. The loss rate for populations of the same species is not estimable with current data, but it is expected to be higher than the loss rate for the species as a whole. With such rapid and massive losses, the resilience of forest ecosystems is decreased, as

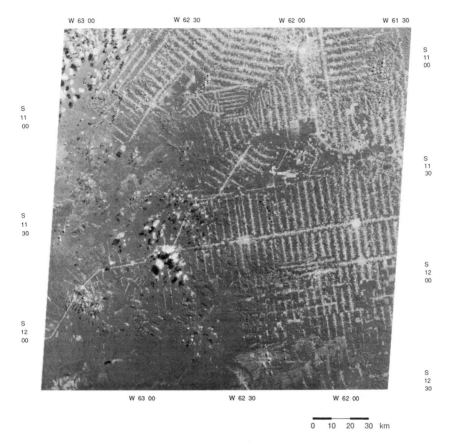

at the end of the road is cloud cover. The image on the right shows the same area on June 16, 1988, with the addition of feeder roads and cleared land. Credit: Instituto de Pesquisas Espaciais (INPE), São José Dos Campos, Brazil.

is their ability to respond to future environmental stresses. Management interventions in the remaining forests, even those naturally reproduced, will be affected by and will further affect the structure of genetic variation of any future forests.

BASIC INVENTORY DATA

One of the most basic problems facing conservation and management efforts for forest species is inadequate and uneven inventory data on the distribution and abundance of trees. In Great Britain, for example, every plant species has been mapped on a computerized 10-km^2 grid system, but tropical areas have not been mapped on less detailed, 100-km^2 grids (Prance, 1984). Many forest tree species, especially in the humid tropics, remain undescribed. As a result, there is no complete global inventory of genetic resources, even at the species level.

The cumbersome and elusive nature of a complete forest inventory (flora) can be illustrated by two well-studied sites. Barro Colorado Island (BCI) in Panama (1,560 ha) and La Selva biological field station of the Organization for Tropical Studies in Costa Rica (1,400 ha) are among the biologically best-known tropical rain forests. The flora of BCI took almost 10 years to complete (Croat, 1978). The work toward a complete flora of La Selva was initiated in the early 1980s, and it is anticipated to be the early 1990s before the flora is published. Despite the fact that the species in both of these areas have been thoroughly collected over the past 25 years, new species are continually being discovered. For example, 15 new species have been found at BCI since the publication of Croat's flora (Gentry, 1986).

CAUSES OF FOREST TREE LOSS

The Eighteenth World Congress (in 1986) of the International Union of Forestry Research Organizations (IUFRO) recognized two major forest-related threats in this century: atmospheric pollution and deforestation. A third threat is the narrowing of the genetic base as a result of commercial forestry operations.

Pollution

In the temperate regions, pollution from industrial and transportation activities, combined with other stresses, is causing severe decline in forest vigor (Scholz et al., 1989). The most severe effects appear to be occurring in Germany and Czechoslovakia, but trees are dying prematurely in most industrialized countries and in some developing countries. Research efforts are under way to identify the causal agents and biochemical pathways through which soil, water, and the trees themselves are affected by pollution. In Germany alone more than 1,000 government-sponsored projects are examining the effects of pollution on forests. The Executive Board of the IUFRO has established a Special Task Force on Pollution to coordinate research and information on this problem, and reviews of the state of knowledge were a major part of its World Congress in Montreal during 1990. Additionally concern has been rising about the consequences to the world's forests of potential global warming attending the destruction of the earth's ozone layer.

Large-scale clearing of tropical rain forest in the Amazon Basin, Goias, Brazil, near the Rio Tocantins, was photographed in 1985. Cleared rectangles are plantations where natural woodland was rapidly being converted to pastureland for cattle ranching and agricultural land for cash crops, such as maize and sugarcane. Smoke rising from a fire burning crop residue or brush appears in the center of the photograph. Credit: National Aeronautics and Space Administration.

Deforestation

Although evidence exists of the effects of pollution on both temperate and tropical forests, deforestation is recognized as the major threat in the tropics. Deforestation is associated with the following activities (Evans, 1982):

- clearance for stationary agriculture,
- producing firewood and charcoal,
- shifting (rotating) cultivation,
- logging,
- expanding urban and industrial areas,

- overgrazing and fodder collection,
- burning (accidental or deliberate), and
- warfare.

The first four activities are the most significant from a global perspective and are discussed here. Political and economic policies for such issues as land tenure and rural poverty also have significant effects on the degree to which lands are deforested.

Clearing for stationary agriculture is occurring in all forest types, but it is most extensive in moist tropical forests. Tropical forests are being cleared for planting industrial perennial crops, such as oil palm and rubber trees, and for cattle grazing in South America and the Pacific islands. African savannah woodland is also being cleared to create small mixed-use farms. Intensive logging is occurring throughout many countries and, as the value of fine hardwood timbers increases, more remote forests are becoming economically attractive for logging. Training in and development of methods for the nondestructive harvesting of trees are insufficient to inhibit the steady destruction of the forest ecosystem.

The production of firewood and charcoal, which at one time was environmentally benign, under new systems or with higher population pressures is now destroying closed and open woodland in drier regions, for example, and is contributing to desertification. Cultivation shifts in Asia encompass 100 million ha annually, mostly on such short rotations that the entire system is threatened. The slash-and-burn agriculture of many parts of Africa, Asia, and Latin America was traditionally sustainable, but current practices have made it another source of loss (Peters and Neuenschwander, 1988). Putting deforested cropland back into production before essential soil nutrients and structures have been replenished also destroys potential forest regeneration. In North America, for example, conversion of forests to new forest types has occurred over large areas when the species of well-established (climax) forests have been replaced by other, pioneer species following clear-cutting.

Narrowing of Genetic Base

Another cause of genetic loss is the possible narrowed genetic base resulting from the intensive breeding of a few economically important species without concern for conservation. The establishment of tree plantations from a few selected individuals, for example, gives the forester an opportunity to exercise much stricter control over the genetic composition of the forest. This leads to a progression, as with modern

An oak savannah in Sirt-Eruh, Turkey, has been partially harvested for fuel
production. The logs in the foreground will be made into charcoal and sold in
urban areas. Credit: Calvin R. Sperling.

crop plants, from the use of wild populations to the use of more
advanced populations in which gene frequencies have been changed to
meet specific requirements and in which genetic variation may be
drastically reduced. As with agricultural crops, breeders tend to maximize the adaptation of these new populations to large-scale plantation
conditions by intensive selection for certain traits, often at the expense
of genetic flexibility and the potential for future adaptive change. It is
possible to provide genetic flexibility by breeding differently enhanced
populations for alternative future uses and to maintain variability in
assorted types of tree populations, but very few tree breeding programs
are designed to do so. (See the multiple-population breeding strategy
described by Namkoong et al., 1980.) Unfortunately, some genetic losses
cannot be recouped because current production is often not backed up
by the holding of extensive off-site collections of tested trees or by major
efforts to conserve variation in natural stands.

WHAT CAN BE DONE?

A species or a population sample of a particular part of its genetic
variation can be conserved through in situ or ex situ conservation. In

situ conservation preserves both the population and the evolutionary processes that enable the population to adapt. Ex situ conservation preserves the genetic diversity extant in the population in a manner that makes samples of the preserved material readily available.

In situ conservation takes place on the site where the trees or their immediate parents are growing naturally. Thus, maintenance of a mature natural forest in a state undisturbed by managerial or other human intervention is conservation in situ, provided the trees are capable of reproducing. Allowing natural regeneration to occur in the same forest

---

WHAT ARE TREE GENETIC RESOURCES?

At a 109-ha nursery operated by the Washington State Department of Natural Resources, 2-year-old Douglas fir seedlings are harvested for planting. The state owns 810,000 ha of forestland, which is managed to produce income for schools. The nursery annually distributes 12 million seedlings. Credit: James P. Blair ©National Geographic Society.

The genetic composition of plants determines to a large extent their structure, physiology, function, form, and life span. A tree's genes can enable it to withstand harsh environmental conditions and attacks by insects or pathogens or to grow faster, straighter, or larger than others around it. The scientist faced with the challenge to develop trees that are suited to particular needs, such as straight-growing trees for lumber manufacture, selects trees that possess the desired gene for breeding programs.

If the need is for readily observable characteristics, such as plant form, it is relatively simple to select trees with the right genes. For the most part, however, the forest scientist selects trees to be conserved without a complete knowledge of what the future will require. In this case, the challenge is to collect a representative sample of known and unknown genes. Such a collection for a particular tree species forms a genetic resource for that species. From this pool, the breeder can select trees with particular genetic traits for direct use in production or as parents in a breeding program to develop improved trees.

is also conservation in situ. Unlike the use of this term for agricultural crops, maintaining planted tree stands in a region where the species occur naturally is also considered in situ conservation. Thus, protection of artificial regeneration that results from sowing of seeds or planting, provided the materials are collected from the same areas where the planting takes place, is also included under in situ conservation.

Ex situ conservation, in contrast, implies that material is protected at a place outside the distribution of the parent population. It may be applied to reproductive material, such as seeds preserved in a seed bank, or to trees planted in arboreta, botanical gardens, or test or conservation stands away from the site of the parent population. Ex situ and in situ conservation are complementary strategies. Ex situ conservation is used where in situ conservation, for various reasons, is impractical or too expensive to maintain. This is the case, for example, when tree populations are under strong demographic or other pressure and their long-term conservation in situ is thereby impossible to secure, or when the germplasm is used in places that are remote from their original locations. (The circumstances for employing both of these conservation approaches are examined in Chapter 4.)

Conservation and use of tree germplasm face technical impediments that must be solved. Trees are long-lived organisms that require large spaces for survival to maturity and for maximum productivity. Techniques for use and conservation, such as establishing biosphere reserves, are known in principle, but little attention has been paid to the genetic management of reserves in which species must be conserved in situ (Palmberg and Esquinas-Alcázar, 1990).

The reproductive biology and genetic structure of tree species do not conform to any single paradigm. Breeding systems vary among species, and many species exhibit mixed mating strategies. Some species flower and set fruit in irregular patterns, which makes their outcrossing (interbreeding) rates unpredictable. The resulting diversity of genetic structure in a population means that different sampling patterns must be used to ensure conservation.

The breeding cycle for trees is currently an order of magnitude longer than that for most agricultural crops and must be accommodated in effective conservation programs. Research is needed to develop tools for rapid or early reproduction of trees. Because forestry must also deal with species of little current commercial importance, methods other than those for managing trees of the same age (even-age plantation management) will have to be considered. These may include partial replacement with younger trees, direct seeding into conservation stands, and control of the size and location of areas to be harvested in the stands. While the methods for conserving trees can draw on many of the techniques developed for agricultural crops, they may need to be tailored to meet constraints, such as long generation times, that are peculiar to most tree species.

RECOMMENDATIONS

It is clear that to maximize the future global utility of forest tree resources, a great deal of basic scientific information and research are urgently needed.

COMPARING TREES WITH AGRICULTURAL CROPS

In addition to differences in size and the space required for growing, trees differ from current agricultural crops in three main respects: known genetic variation, socioeconomic value, and multiplicity of uses.

Known Genetic Variation

Forest trees can be inbreeders (sexual reproduction is through self-pollination and, hence, the trees are genetically very similar), outbreeders (reproduction through crossing of genetically different trees and, hence, the trees are more genetically heterogeneous), or intermediate (use of both strategies to varying degrees). Thus, different sampling patterns are required for ex situ conservation. Most tree species are at a very primitive stage of genetic selection from wild types compared with agricultural crops, some of which have passed through thousands of generations of domestication and up to perhaps a hundred generations of selective breeding. With the exception

In Burkina Faso, a country in the drought-prone Sahelian region of west Africa, researchers cultivate a fast-growing species, Leucaena leucoephala. *The tree has been planted throughout the world for use as fuelwood and forage, and to help stabilize deforested or semiarid ecosystems. Credit: U.S. Agency for International Development.*

More comprehensive study should be made of the effects of global climatic change and pollution on forest tree species.
Much attention has been focused on the potential consequences of global climate changes brought about by increased concentrations of heat-absorbing greenhouse gases, such as carbon dioxide, methane, nitrous oxide, and chlorofluorocarbons. Some experts suggest that the greatest effects of increases in atmospheric carbon dioxide, mostly resulting from the burning of fossil fuels over the past 100 years, will occur at higher elevations and more northern latitudes. These speculations, however, remain to be substantiated by scientific study (Silver,

---

of programs for a few woody species (e.g., willows, eucalypts, pines, and poplars), few tree breeding programs have progressed beyond the second or third selective cycle.

Socioeconomic Status

The fact that forests evolve and that regeneration is susceptible to positive and negative human influence is often not understood. (It is only when forest products are in short supply or lacking that people realize the need for them.) Conversely, it has been recognized for 10,000 years that food requirements can be produced or enhanced by the deliberate creation and management of resources. Wood itself is a low-value and high-volume product, and food will always have survival priority over wood products or other forest benefits.

Quantifying the value of benefits derived from trees is difficult, for several reasons. First, many forests used today originated naturally, so no costs of establishment or stumpage (market) values have been applied in economic analyses. Second, some benefits, such as soil holding or improvement, or water flow moderation, have importance to society that is beyond their obvious environmental advantages. It is, however, difficult to determine such social values. Third, rotations of planted tree crops normally extend over many years, which creates difficulties in assigning variable discount and inflation rates. Given high-establishment costs in the early years and low-product values in later years, forestry projects always appear poor investments unless their social benefits are taken into account. The same principle applies to the allocation of funds for conservation of genetic resources.

Multiplicity of Uses

There is great international interest in so-called multipurpose trees (Burley and von Carlowitz, 1984), but in practice virtually all tree species can be used for more than one purpose. In this respect, they differ from many major agricultural species, which are generally grown for single products and often for specialized uses.

Land in the foreground was cut for lumber up to the boundary of the Gifford Pinchot National Forest and replanted with younger trees. Clear-cut patches in the national forest are visible. In Washington State, private enterprises are required to replant logged areas within 3 years. Credit: James P. Blair ©National Geographic Society.

1990). Forests themselves are both sources and consumers of atmospheric carbon dioxide. The extent to which existing or newly planted forests could affect the atmospheric concentration of this gas remains a subject for scientific investigation and debate.

More general agreement exists about the link between forest decline and pollution. However, it is difficult to assess precisely the environmental and physiological pathways that are affected by pollutants. Research is under way to elucidate and document the effects of acid rain and other atmospheric pollutants on forest trees. Researchers must look at not only the possibility of forest loss, but also at any decline in the genetic diversity within forest tree species.

Increased efforts are needed to provide an accurate inventory of forest trees.
The pace at which inventories are being undertaken or completed is not commensurate with the rate at which the subtropical and tropical vegetation is being lost. In western Ecuador, for example, 100 new

species described during the preparation of the flora of the Rio Palenque field station in Los Rios Province are now found only in the field station, which covers an area of less than 1 km^2 (Gentry, 1986). Greatly improved data on inventory are needed, particularly for tropical and subtropical regions. A plan for developing an inventory of the world's forests must be developed and implemented.

2

Multiple Uses of Forest Trees

Many agricultural species are grown for single products and for specialized uses, but forest trees are often grown for more than just pulp or timber, and even those products have multiple end uses. Because genetic variability has been demonstrated in most of the growth, morphological, and anatomical properties of trees, the economic values for which trees are managed can depend on their genetic variability. It also seems clear that the wide range of management systems and intensities that have already been applied to trees and forests have some influence on their genetic variation.

BENEFITS OF TREES AND FORESTS

Forests and trees provide a variety of benefits to humans. Those benefits can be divided into two major categories: direct and indirect benefits (Burley, 1987).

Direct Benefits

Direct benefits and uses include forest products of economic importance, such as sawtimber and numerous other construction materials, fodder, and fuelwood. Trees have been used throughout the world for millennia, and today over one-half of the wood used each year is burned for heating and cooking. In some countries in Africa, for example, 70 percent or more of the total energy used is provided by wood (Table 2-1).

TABLE 2-1   Share of Total Energy Use Provided by
Wood, Selected Countries, Early 1980s

Country	Wood Share of Total Energy Use (percent)
Africa	
Kenya	71
Sudan	74
Nigeria	82
Tanzania	92
Malawi	93
Burkina Faso	96
Asia	
China	>25[a]
India	33
Indonesia	50
Nepal	94
Latin America	
Brazil	20
Costa Rica	33
Nicaragua	50
Paraguay	64

[a] Includes agricultural wastes and dung in addition to wood and charcoal.

SOURCE: S. Postel and L. Heise. 1988. Reforesting the Earth. Worldwatch Paper 83. Washington, D.C.: Worldwatch Institute. Reprinted with permission.

Current global wood production approximates 1.5 billion m³ annually (Table 2-2), as recorded in national government statistics, but the use of additional vast quantities of wood goes unrecorded. Another group of wood uses comprises reconstituted and reassembled wood, including veneers, chipboard, fiberboard, pulp, and paper. Total global trade in paper products approximates $60 billion of imports and exports (Food and Agriculture Organization, 1985b).

The harvesting of plant products provides income to national economies and individuals. This type of employment is often seasonal and provides off-farm income when seasonal on-farm workloads are low. The harvesting, sale, and processing of products from trees in combination with farming also reduce the risks associated with incomes based on one or two cash crops.

The number of products provided by trees worldwide is extensive. The wood, bark, leaves, fruit, seeds, and roots of trees yield food, fodder, shelter, medicine, fiber, resins, oils, and numerous other products used for subsistence and industrial purposes. In some countries the products from trees are important contributors to individual, village, and national economies (Myers, 1983a). Forest products can serve as feedstocks to support a wide range of local and commercial industries, drugs, food for human consumption, and fodder for animals (Panday, 1982; Parkash and Hocking, 1986). In India, Myers (1988) estimates the amount of annual revenue derived from such minor forest products is $200 million. This equals or exceeds India's revenue from wood extracted from the forests.

Management of forests for this variety of uses ranges from intensive cultivation on large industrial tracts, such as in the southeastern United States, to purely exploitative extraction with no plans for forest regeneration. In some societies, nuts, leaves, bark, roots, latex, and various other parts of trees are collected as semidomesticated agronomic crops. In the western Amazon, for example, tapping rubber and collecting Brazil nuts are combined in forests, known as extractive reserves, that may provide income while conserving the tropical forest ecosystem (Cowell, 1990; Pearce, 1990).

TABLE 2-2 World Production of Industrial Roundwood, 1985[a]

Country	Volume (million m³)	Share of Total (percent)
United States	347	23
Soviet Union	275	18
Canada	165	11
China/Taiwan	93	6
Brazil	58	4
Sweden	49	3
Finland	39	3
Japan	33	2
Malaysia	32	2
France	29	2
All others	383	26
World total	1,503	100

[a] Industrial roundwood is defined as logs, pulp, and other raw materials used to manufacture wood products.

SOURCE: S. Postel and L. Heise. 1988. Reforesting the Earth. Worldwatch Paper 83. Washington, D.C.: Worldwatch Institute. Reprinted with permission.

A Sherpa woman in the central highland region of Nepal carries fodder from woodland on a nearby hillside to feed livestock. Fodder can be an important part of household income, but its overcollection in areas of high population density can cause serious damage to nearby forests. Credit: Steven King.

Indirect Benefits

Indirect benefits of trees include environmental protection and amenity (social) values. Trees contribute to the sustainability of land productivity by contributing to the formation, structure, and fertility of soil in many ways. For example, they form symbiotic relationships with nutrient-absorbing fungi or nitrogen-fixing bacteria and fungi, and moderate water flow and loss while binding soils to prevent erosion. Trees also contribute significantly to reducing greenhouse gases and associated global temperature rises by converting carbon dioxide to carbohydrate through photosynthesis and then locking it up within their structures.

The indirect costs of ignoring the protection provided to the ecosystem by healthy forests can be very high. Deforestation and poor cultivation practices have been cited as major causes for dramatic rises in sediments in river systems around the world (Postel, 1989). Increased sedimentation into reservoirs as a result of soil runoff is dramatically reducing the life span of several hydroelectric dams. In The Philippines, it is estimated that sedimentation rates in two reservoirs increased by more than 100 percent between 1967 and 1980. As a result, the life span of the Philippine Ambuklao Dam project has been cut in half. A similar process is occurring in Costa Rica, where siltation may significantly reduce the lifetime energy production of the Cachi hydroelectric project (Postel and Heise, 1988).

Forests are much more than an assemblage of woody trunks with leafy canopies. Trees determine the structure and organization of forest ecosystems, which provide habitat for a multitude of species of plants, animals, and microorganisms used directly and indirectly by humans. Species diversity itself may be considered an indirect product of forests, especially in the tropical moist forests, which contain at least one-half of all known plant and animal species (Food and Agriculture Organization, 1985c).

The amenity benefits of forests and trees include scenic beauty, the opportunity to view wildlife, and access to wilderness areas for study and recreation. It is widely accepted in developed countries that such amenity resources should be made available by the government for the benefit of the people, possibly with direct user fees as well as with indirect contributions through taxes. In developing countries, governments are realizing, however, the importance of natural resources, particularly forests and wildlife, in attracting tourists and foreign exchange. Trees also serve as cultural and religious symbols. One example of this is the sacred forests and tree species that are found in India and other countries in Asia.

LAND USE SYSTEMS INVOLVING TREES

Land use systems form part of the full spectrum of uses of forest trees. A variety of land use systems incorporate either deliberately planted or naturally occurring trees. These systems can be divided into three major types: natural vegetation management, agroforestry, and industrial plantation forestry, each of which produces major benefits. The first type of system does not usually involve selection of trees with a specific genetic composition (genotypes) or any other direct management of genetic variability, whereas the second and third types use mixtures of genotypic selection (and occasionally direct breeding) and indirect genetic management.

Natural Vegetation Management

Conservation, protection, or management of naturally occurring forests to preserve their biotic integrity is natural vegetation management.

Trees are a source of fuel throughout the world. Kilns near Ilheus, Bahia, Brazil, produce charcoal from discarded saw timber for cooking fuel and other uses. Credit: Douglas Daly.

This can include selective harvest of forest products. Considerable research has been undertaken on the sustainable management of tropical moist forests (Mergen and Vincent, 1987; Wyatt-Smith, 1987), and it is conservation of these forests that is receiving the most attention in the media. The genetic implications of such natural vegetation management systems remain unexplored, yet the population sizes and the structure of genetic variation maintained by those systems would undoubtedly affect the viability of these ecosystems. Relatively little attention has been given to the conservation of trees and shrubs in drier zones, although they may be the sole source of some plant materials, especially for fuel and fodder. Often, too, the dry zones contain tree species that could have great potential as exotic (nonnative) trees for other regions.

Agroforestry

A range of land management systems, including rural development forestry, alley farming, and silvipastoral systems, are often collectively termed agroforestry. Trees, crops, and domestic animals are mixed simultaneously or sequentially in such systems to obtain increased and sustained multiple benefits from limited land resources. In rural development forestry, planted trees are used for the benefit of rural communities or individual farmers. Such systems include the small plantations of community and farm woodlots and the establishment of trees in farmland for shade, shelter, fuelwood, soil improvement, or other purposes. The selection of source populations and species for these systems is just beginning.

One of the most widely researched agroforestry systems is alley farming. This is a term for a form of land use in which one or more rows of trees are planted alternately with several rows of agricultural crop plants. The trees in this system are hedged (cut back frequently) and contribute decomposing leaves (mulch). This practice increases soil fertility and provides fuelwood as well as other benefits. Despite considerable research, however, adoption rates for alley farming remain low.

Silvipastoral systems incorporate tree and shrub management and animal husbandry. The trees are used for fodder production, shade, and pasture improvement. The intensity of such operations varies from extensive range management in dry zones to intensive trees-over-pasture systems in areas of higher rainfall. The trees may be planted or occur naturally.

Research on agroforestry systems is increasing and often benefits from a multidisciplinary approach. Much of the work is being conducted as part of the research program of the International Council for Research in Agroforestry. Related research and development work on multipurpose trees and agroforestry systems in India and in southwest and southeast Asia is being conducted as part of the Forestry/Fuelwood Research Education and Development project coordinated by the Forestry Research Institute of Malaysia (Plucknett et al., 1990).

Industrial Plantation Forestry

Industrial plantation forestry is practiced on large areas that are established and managed intensively, often with exotic species, for the production of timber to supply sawmills, pulpmills, veneer factories, chipboard plants, and so on. In developing countries, the plantations are usually owned and managed by state enterprises, although community, company, and private ownerships of forests exist in the tropics.

In temperate countries, industrial plantations contain largely conifers (especially larch, pine, and spruce) or fine hardwoods (ash, beech, and oak). In tropical and Mediterranean countries, plantations consist of

TABLE 2-3   Growth Rates of Selected Tropical Plantation Tree Species

Plantation Development	Species	Yield (m³ha⁻¹a⁻¹)ᵃ	Rotation (years)
Scott Paper Co., Costa Rica	*Pinus caribaea*	40	8
Aracruz Florestal, Brazil	*Eucalyptus grandis*	35	7
Jari Florestal, Brazil	*Gmelina arborea*	35	10
Jari Florestal, Brazil	*P. caribaea*	27	16
Fiji Pine Commission	*P. caribaea*	21	12–15
Seaqaqa plantations, Fiji	*Swietenia macrophylla*	14	30
Viphya Pulpwood Project, Malawi	*P. patula*	18	16
Commonwealth New Guinea Timbers, Papua New Guinea	*Araucaria* spp.	20	40
Paper Industries Corp. of the Philippines	*Albizzia falcataria*	28	10
Shiselweni Forestry, Swaziland	*E. grandis*	18	9
Usutu Forest, Swaziland	*P. patula*	19	15

ᵃ Cubic meters per hectare per annum.

SOURCE: J. Evans. 1982. Plantation Forestry in the Tropics. New York: Oxford University Press. Reprinted by permission of Oxford University Press.

mainly eucalypts, *Gmelina* (widely used for timber), mahoganies, pines, and teak, although many other species have potential for specific conditions, such as laurel and *Leucaena*. The area of plantations (both industrial and those not strictly for industrial purposes) planted in the tropics totals 1.5 million ha annually with a current cumulative standing area of approximately 11 million ha.

Growth rates for tropical plantation species are shown in Table 2-3. By comparison, the mean annual growth of a naturally occurring climax tropical forest is rarely greater than 3 to 4 m³ per hectare.

Constraints and Opportunities of Land Use Systems

Although managed breeding programs are used in some areas, intensive forestry on a large scale is possible only in relatively simple ecosystems. For large-scale but low-intensity management, as for amenity, fuelwood, or protection forests (e.g., windbreaks), only low-cost breeding and regeneration programs are economically feasible for maintaining forests or afforesting impoverished landscapes. For small-scale, low-intensity forestry, as practiced to maintain ecosystems, feasible management could still include controlling the size and distribution of clearings or reproduction patches. For more intensively managed areas in which large-scale afforestation or reforestation is needed, as in some tropical areas, the management technology may still require research and development.

Current investment levels in land use systems are rarely high enough to duplicate the intensive breeding or conservation procedures used with agronomic crops, even for species currently used in industrial forestry. Even if high-genetic gains in productivity can be expected, investment is generally low. This may be due to the length of time required for forestry to provide investment returns and for genetic techniques to influence forest productivity. It may also be due to the misconception that nonindustrial uses are of limited value. Investors find little likelihood of immediate profit even though the long-term, societal benefits may be high.

Research to develop faster breeding techniques and low-cost forestation programs for using selected stocks of many different species could induce more investment. Even when the economic use and value of forestlands can be feasibly and substantially enhanced by appropriate breeding, however, economic inhibitions continue to deter private sources from developing the genetic resource.

INCREASING THE USEFULNESS OF TREE GENETIC RESOURCES

Given the lack of knowledge about the nature and distribution of forest genetic resources, substantial efforts are needed in two other traditional stages of using plant genetic resources: plant exploration and evaluation. Both are necessary before sustained utilization of the resources can be assured. In addition, efforts should be organized to coordinate the exploration and evaluation of programs of various agencies and governments. Finally, programs are needed to test and conserve species whose potential value is not yet known.

Exploration

For a given species, its natural range must be explored and areas to which it has been introduced must be identified. Determination of its taxonomic status (relationship to other species) and study of its natural breeding system are also essential. Such information is reasonably well known for most of the temperate-zone species among the 100 tree species used in commercial plantations. From several hundred to a few thousand other potentially valuable species, particularly in the tropics, are not yet widely used either for commercial purposes or for rural

Medicinal quinine for the treatment of malaria was originally derived from the bark of the cinchona, a tree native to the South American rain forest. Credit: U.S. Forest Service.

development. Little is known about their attributes, status, or distribution (Burley and von Carlowitz, 1984). For many, their reproductive system, population variation, and distribution are also little understood.

Evaluation

Ex situ field trials coupled with molecular or biochemical analyses are needed to determine the pattern and extent of genetic variations for many species. Well-designed, replicated experiments on an array of field sites provide data on the variation among populations in survival, growth, productivity, and qualitative characteristics. Through provenance testing (repeated trials on a range of sites), estimates can be made of the importance of genotype-environment interaction effects. This, in turn, indicates the extent to which individual populations should be conserved separately and also bred separately in the future. (Details of the design, management, and assessment of such trials are given in Burley and Wood [1976] and, for multipurpose trees, in Burley and Wood [1987].)

Coordination

When a species is indigenous to several countries and is useful in several others, great advantages accrue from organizing coordinated programs of exploration, collection, and evaluation. Such coordination would minimize the number of collecting missions and maximize both the comparability of sampling within the natural range and of testing in experimental locations. Often such projects have been undertaken by organizations in developed countries. Examples include the Commonwealth Scientific and Industrial Research Organization in Australia (many eucalypts), Centre Technique Forestier Tropical in France (various tropical species), the Danish Forest Seed Center (*Gmelina*, teak, some tropical pines), and the Oxford Forestry Institute in the United Kingdom (Central American pines and hardwoods, some Asian tropical pines, and some indigenous African acacias).

The Food and Agriculture Organization of the United Nations, in a program for arid-zone species supported in part by the International Board for Plant Genetic Resources, took a somewhat less centralized approach by asking research institutions within the natural-range countries to make collections and pool them for distribution to institutions interested in evaluating them. In either system, herbarium specimens are collected for taxonomic study, and preliminary observations of breeding systems can also be made during the exploratory phase.

Development of Programs for Untested Species

For species that are without clear potential production value and that are not adequately included within ecosystem reserve areas, no programs to conserve genetic resources exist. Species that might serve vital functions for ecosystem productivity, recreation, or other diffuse values, but that are not included in protected areas, have no constituency to encourage investment in their management or conservation. The identification and management of such populations for their genetic resources have received increasing attention in conservation efforts (Schoenwald-Cox et al., 1983; Wilson, 1988).

RECOMMENDATIONS

Industrial production forestry and the other land use systems described in this chapter are evolving in response to the changing needs of people, the environment, and industry. To maintain the widest

In Honduras, a farmer taps a pine tree to collect resin. The establishment of extractive reserves is proposed by some experts as a way to allow income-producing resources to be collected while preserving trees. Credit: Food and Agriculture Organization.

genetic array of options for the future, it is necessary to understand and work with the existing structure of genetic variation in trees.

Additional cooperative efforts among nations are needed to develop coordinated programs for exploring, collecting, and evaluating tree genetic resources.

Cooperation among nations or between national and international programs can make efficient use of limited funds and technical capacities to promote exploration and evaluation. Exploration is necessary to determine the nature and extent of genetic diversity for forest tree species. Data on the geographic distribution, taxonomy, biology, and ecology of a species may be needed to supplement knowledge obtained from material already in reserves or collections. The information gathered through evaluation is crucial for making wise decisions about conserving and using species and populations.

For many species that originate in a country or region where they may have relatively little value but may be of major economic significance elsewhere, three questions are raised: (1) Who should conduct and pay for exploration and any subsequent in situ conservation? (2) Who should conduct and pay for testing and establishing ex situ conservation stands? (3) Should the country of origin restrict export or charge origin fees for its genetic material? These issues have yet to be raised seriously in forestry, but they are undergoing major international debate for food crop species.

Forest genetic resources programs should conserve species that lack clear present or potential value and those that have known potential value.

Breeding programs that do not rely on expensive testing or lengthy regeneration techniques should be instituted for a large number of species, especially those being newly brought into genetic management programs. For the many hundreds of species of potential value, genetic surveys and sampling for multiple populations are needed and might be combined with preliminary breeding operations. Of the species currently being tested or considered for multiple breeding populations, such as teak in Thailand (Wellendorf and Kaosa-Ard, 1988), most would require an increase of 5 to 10 times the current funding. To develop the species of known potential value, an order-of-magnitude increase in the number of species included in exploration and testing programs is needed. To conserve the species that lack clear current or potential value, additional efforts are needed.

3

Structure of Genetic Variation

One of the most fundamental requirements for improved management of forest tree diversity is understanding the biological dynamics of genetic variation within and between tree species. Information about the genetic architecture of tree populations will make it possible to develop sounder strategies for their conservation and use.

MATING SYSTEMS AND GENE FLOW

Considerable variation exists among tree species with respect to the extent of genetic diversity and the way such diversity is spatially or temporally organized within and among populations. The extent and the pattern of genetic diversity in forest trees are strongly regulated by their mating patterns and gene flow. Despite decades of research on the reproductive biology of forest trees, knowledge about the subject is still limited.

Mating Mechanisms

Mating systems among trees are quite varied. Most tropical forest trees have hermaphroditic flowers; that is, the flowers contain male and female reproductive organs (Ashton, 1969; Bawa, 1974, 1979). In contrast, a vast majority of temperate tree species are monoecious; that is, they are characterized by the presence of unisexual flowers, but male and female flowers occur on the same plant. Dioecy (the presence of separate male and female plants) occurs in temperate and tropical species, but

51

it is more common in the latter (Bawa and Opler, 1975). The evolutionary dynamics of forest trees is further complicated by the existence of mixed mating and reproductive systems. Selfing (self-pollination or self-fertilization) is possible in hermaphroditic and monoecious species, but it is usually prevented by a wide variety of genetic mechanisms or differences in the maturation of male and female floral parts.

Many tropical species are genetically self-incompatible; that is, little or no seed is set following self-pollination (Bawa, 1974; Bawa et al., 1985). Surprisingly little is known about self-incompatibility in temperate species with hermaphroditic flowers; in the monoecious species, two mechanisms prevail that prevent or reduce selfing. First, the maturation of male and female flowers at different times, as happens, for example, in most temperate coniferous trees, reduces the possibility of selfing. Second, self-sterility, which also occurs in most conifers, causes most selfed seeds to be aborted before they mature. This mortality is assumed to be due to lethal recessive genes that are brought together when individuals that are closely related genetically are crossed (inbreeding).

In sum, designing tree reserves and managing natural or artificial stands of trees requires an understanding of the reproduction of the trees involved. Without it, tree populations could, for example, fail to reproduce, or they could experience excessive or unintended inbreeding and might thus become endangered.

Inbreeding

Various mechanisms can eliminate the possibility of selfing, but they do not exclude inbreeding. The level of inbreeding is determined not only by the nature of the reproductive system, but also by family structure, which itself is influenced by pollen and seed dispersal characteristics. Dispersal of pollen and seeds over a limited area can increase the genetic relatedness of nearby individuals in a mating population and, thereby, the potential for inbreeding. The longevity of some trees over others in the population can also result in a relatively few individuals genetically dominating the gene pool. This, combined with such factors as asynchronous flowering of male and female flowers, overlapping generations within the populations, and unequal sex ratios, can potentially increase the level of inbreeding. Apomixis, or uniparental reproduction, which appears to be quite common in some tropical trees (Ashton, 1988), may also contribute to inbreeding. The net result of these factors is that the effective population size in terms of reproductive capacity is often much smaller than the total number of adult trees.

Outcrossing

Outcrossing refers to the mating of genetically nonidentical individuals. The rate of outcrossing is often estimated by the use of genetic markers in the form of allozymes (genetic variants of enzymes). In temperate and tropical trees, outcrossing rates have been found to be very high. Although the outcrossing rate can range from 0 (no outcrossing) to 1 (100 percent outcrossing), it varies from 0.60 to 1 in the tree species examined by Knowles et al. (1987), and wide variations in rates occur among individuals even of the same population. The distances between trees and the timing of reproductive flowering (i.e., spatial and temporal variation) can also affect outcrossing rates. Rates in tamarack, for example, have been found to vary with stand density, the outcrossing rate being lower in less dense than in more dense stands. A practical implication of this result is that spatial isolation of trees due to forest decline or degradation may increase the level of inbreeding and its concomitant effects.

Gene Flow

Gene flow implies an exchange of gametes or genes among dispersed trees, and it is inversely related to population differentiation. It occurs through the movement of pollen and seed, which can be dispersed by a wide variety of nonliving and living mechanisms. Knowing how populations are expanded, maintained, or restricted by gene flow is essential for managing distinct tree populations in their original environment.

Gene flow determines the geographic scale over which populations may be differentiated from each other. Population differentiation can also occur in response to selection resulting from local variations in environmental factors. Discriminating between selection effects and migration-induced patterns, however, is difficult. Moreover, selection or sexually divergent migration, which can be frequent among tree species, can have significant impact on gene flow (Gregorius and Namkoong, 1983; Namkoong and Gregorius, 1985). Where lack of gene flow occurs, the potential for future development of distinct populations exists.

All temperate conifers and many hardwood species, such as oaks and poplars, are wind pollinated. Although most pollen is dispersed by windfalls near the paternal tree, much is still carried farther away, several hundred meters or more (Levin and Kerster, 1974). This, in combination with the high densities of these temperate, wind-pollinated species, results in a potentially high frequency of pollen exchange.

Some temperate trees are pollinated by animals (including insects), but virtually nothing is known about the extent to which animal pollinators move among trees. In fact, in most instances even the pollen vectors (carriers) have not been identified. Moreover, in a few instances in which both bird and insect pollinators are effective, as in *Camellia japonica* (Japanese camellia), the relative visitation frequencies of the pollinators are not known, but it is known that the distances over which they distribute pollen can vary considerably.

The diversity of insect and other animal pollinators in the tropics is immense, ranging from bats with a wingspan of 1.5 m to tiny wasps that are 1 or 2 mm in size. Many species, however, are pollinated by medium-sized to large bees (Bawa et al., 1985). Other pollen vectors are moths, beetles, and the generalist insects, which are the wide variety of small bees, beetles, butterflies, wasps, and flies that visit flowers.

THE PACKAGING OF GENETIC INFORMATION

A gene, in the classical sense, is the basic unit of heredity and has one or more specific effects on an organism. Genes are segments of DNA (deoxyribonucleic acid) in the nucleus of the cell, linearly arranged to form threadlike structures called chromosomes. The term "locus" refers to the position of a gene on the chromosome, and it is sometimes used interchangeably with the term "gene" when referring to regions of DNA that are influencing a trait.

Alternate forms of a gene found at the same locus are called alleles. Some genes have many alleles, which allow for multiple gene products and therefore multiple phenotypes. For example, multiple alleles have been identified for many of the genes that code for human blood proteins. The allelic frequency refers to the proportion of loci in the population occupied by each allele. A gene for which there is more than one common allele is called polymorphic.

Most organisms are diploid, meaning they carry two copies of each gene. If both copies are the same allele, the individual is said to be homozygous; if the two copies are different alleles of the gene, the individual is heterozygous for that locus.

DNA is the chemical that carries genetic information in all living organisms. The DNA molecule itself is an elongated double helix, often compared with a long, twisted ladder. Corresponding to the rungs of the ladder are two bases, forming a "base pair." The sequence of base pairs at a given locus confers the specificity required for transmission of

Relatively few tree species are pollinated by bats and birds, although some important exceptions are the durian fruit (*Durio*) of Asia, the timber genus *Caryocar* of the Amazon, many tropical legumes, and species of the Bombacaceae. However, these animals do pollinate a large number of other species in the forest.

The extent to which pollen vectors move pollen between plants largely depends on their foraging behavior and the spatiotemporal distribution of flowers. Insects and other animals may bring about extensive exchange of pollen among individuals scattered over a wide area (Regal, 1977). This is particularly true where specificity (uniqueness) between pollen vector and the plant is high. An example is the genus *Ficus* (fig), in which about 800 species are pollinated by different species of wasps (Wiebes, 1979). The high specificity enables the fig trees to reproduce and persist even in very low densities (Janzen, 1979). However, this extreme specificity between pollen vector and plant is not common among tropical forest trees.

---

information via the genes. Due to variability of the base pairs, the number of different alleles that could theoretically be formed from even a very short piece of DNA is extremely large. For instance, a segment of DNA with only 10 base pairs could have more than 1 million different codes. The "average" gene is thought to have up to a thousand base pairs; thus, the DNA structure provides for a phenomenal amount of variation in the genetic code. The difference between two alleles is often as simple as a substitution of a single base pair, but this may correspond to a significant difference in the phenotype resulting from those alleles.

Variation can readily be observed at several levels: between species, between major types within a species, between populations within a major type, and between individuals. An individual's genetic composition, or genotype, in conjunction with the environment in which that individual is found naturally, determines the phenotype or observable characteristics.

Certain traits are controlled by a single gene and are referred to as qualitative or Mendelian traits. Many characteristics, however, are influenced by a larger number of genes; these are called quantitative or polygenic traits. The cumulative action of these genes influences the expression of the trait, but the effect of any single gene is small and cannot generally be isolated in the phenotype. Important production traits, such as yield, growth rate, and straightness, are examples of polygenic traits. Occasionally, a "major" gene, one that has a stronger influence on the trait, can be detected, but there is nevertheless modification of the trait by other genes with smaller effects.

The fruit of *Protium nitidifolium* has a seed-containing stone encased in a white sweet pulp that is attractive to some birds and bats. Credit: Douglas Daly.

Little is known about the distances over which animal vectors move pollen. Territorial birds, such as hummingbirds, are more sedentary than nonterritorial ones, and consequently, their pollen dispersal is extremely limited (Linhart, 1973). Bats, on the other hand, are known to forage over a range of several kilometers (Heithaus et al., 1975). Similarly, medium-sized to large bees can move pollen among plants that are several kilometers apart (Frankie et al., 1976; Janzen, 1971).

Seeds are also dispersed by a number of mechanisms. The seeds of temperate conifers and many hardwood species, such as maples, poplars, and willows, are dispersed by wind. The effects of such dispersal patterns on limiting the effective population size in larger, continuous populations can be small (Wright, 1962). In nut-bearing trees, squirrels and other rodents are the most common dispersers. Seed dispersal by wind is common in deciduous forests, but in evergreen forests, birds of various sizes and classes disperse the seed of most species. Mammals, especially bats, are another common group of dispersers. Almost no data exist, however, about the distances over which dispersers transport seeds in either the temperate or the tropical zone.

Neutral Alleles and the Quantification of Gene Flow

Alleles are the alternate forms that a gene may take, which may cause selectively important differences to exist among individuals. Neutral alleles provide no selective advantage or disadvantage and, therefore, are useful for modeling gene flow in populations. The flow of neutral alleles between populations can be quantified by estimating the value of Nm (N = effective size of the local population; m = average rate of gene migration by means of pollen and seeds). The principal reason for estimating Nm is that its value has been shown to indicate the relative importance of gene flow and genetic drift in explaining the observed patterns of genetic differentiation in a population (Slatkin, 1987). If Nm is less than 1, then changes in allele frequencies resulting from genetic drift of neutral alleles can occur. Such changes are not likely if Nm is greater than 1.

As measured by electrophoretically distinguishable alleles, which are often considered to be neutral variations, the values for Nm in a range of forest tree species indicate fairly high levels (> 1.0) of gene flow (Table 3-1). This is to be expected because most trees have high outcrossing rates, and there is a positive association between outcrossing rate and the level of gene flow (Govindaraju, 1988a). Wind-pollinated species have higher outcrossing rates than animal-pollinated species (Govindaraju, 1988b). The effects of seed dispersal on levels of gene flow have not yet been quantified (Hamrick and Loveless, 1987).

Nonrandom assemblages of genotypes impose a structure on the

TABLE 3-1   Relation of Levels of Gene Flow (Nm) to Pollination Mechanism in Selected Forest Tree Species

Species	Number of Populations Sampled	Nm^a	Pollination Mechanism[b]
Abies balsamea	4	4.546	W
A. lasiocarpa	3	7.300	W
Bertholletia excelsa	2	1.626	A
Eucalyptus caesia ssp. *caesia*	7	0.001	A
E. caesia ssp. *magna*	6	0.174	A
E. cloesiana	17	1.081	A
E. delegatensis	8	0.606	A
E. obliqua	4	0.438	A

(continued)

TABLE 3-1 (Continued)

Species	Number of Populations Sampled	Nm^a	Pollination Mechanism[b]
E. pauciflora	3	5.444	A
Larix laricina	10	6.810	W
Liriodendron tulipifera	3	1.000	A
Picea abies	10	1.967	W
P. engelmannii	3	1.000	W
P. mariana	21	3.620	W
Pinus attenuata	4	1.031	W
P. banksiana	32	4.270	W
P. contorta ssp. latifolia	9	4.620	W
P. contorta var. latifolia	5	7.840	W
P. jeffreyii	14	1.350	W
P. longaeva	5	4.050	W
P. monticola	28	1.340	W
P. muricata	7	0.651	W
P. nigra ssp. austriaca	6	2.983	W
P. nigra ssp. dalmatica	5	1.080	W
P. ponderosa	6	2.926	W
P. radiata	5	1.070	W
P. rigida	11	8.776	W
P. sylvestris	14	7.483	W
P. taeda	2	2.780	W
P. virginiana	2	2.341	W
Populus deltoides	10	2.962	W
P. trichocarpa	10	3.020	W
Pseudotsuga menziesii	6	3.041	W
Quercus coccinea	3	2.025	W
Q. ilicifolia	3	2.470	W
Q. marilandica	3	0.579	W
Q. palustris	2	3.614	W
Q. rubra	2	0.805	W
Q. velutina	3	0.540	W
Sequoiadendron giganteum	34	2.190	W

[a] N = effective size of local population; m = average rate of gene migration by means of pollen and seed. Values of Nm less than 1 indicate the potential for genetic drift.
[b] W = wind pollinated; A = pollinated by an insect or other animal.

SOURCE: Adapted from D. R. Govindaraju. 1988b. Relationship between dispersal ability and levels of gene flow in plants. Oikos 52:31–35. Reprinted with permission.

population, which can be measured as the departure from standard statistical (Hardy-Weinberg) proportions. Structure can result from selection and mating among closely related individuals (e.g., siblings or cross-generational relatives) that are near one another. In studies of *Pinus sylvestris* (Scots pine) (Tigerstedt, 1984) and *Liriodendron tulipifera* (tulip tree) (Brotschol et al., 1986), some evidence exists for such inbreeding or mating within localized populations. Thus, although the potential exists for both wide intermating and for neutral alleles to be randomly distributed, there is growing evidence that various genotypes in plant populations are not randomly distributed.

Nonrandom mating within populations of continuous forest stands has been reported in several cases: *Pinus ponderosa* (western yellow or ponderosa pine) (Linhart et al., 1981a,b), *Cryptomeria* (red cedar) (Sakei and Park, 1971), *Eucalyptus* (eucalypts) (Moran and Hopper, 1987), and in some tropical species (O'Malley and Bawa, 1987). A well-documented example of genetic structuring is a small population of *Pinus taeda* (loblolly pine). Allelic differences among groups have been reported in the continuous population and are assumed to be due to three separate regeneration episodes that established the current population (Roberds and Conkle, 1984). Thus, genetic variations within and among populations can be differentially distributed, and sampling or collecting that variation must include those structural variations and not depend on random methods.

ESTIMATING GENETIC VARIATION

Establishing priorities for the conservation, management, and use of tree genetic resources requires an understanding of the degree of diversity among and between specific populations. The scientific methods used to distinguish levels of variation are the working tools for shaping decisions on which resources should be devoted to managing distinct tree characteristics. Two approaches have been used to study genetic diversity in tree populations: provenance testing and allozyme screening. The two approaches are used with different objectives in mind, and the results obtained from each have different applications in theory and practice.

Provenance (geographic origin) testing is performed by gathering seeds from different populations and observing variation in performance (e.g., height, diameter, color, yield) in plants grown under uniform environmental conditions within one or more planting sites. The populations sampled are generally from different geographic and climatic regions. This approach constitutes the basis of provenance testing in forestry. Allozyme screening is based on the survey of genetic diversity as revealed by variation in enzymes at specific gene loci. In general, both provenance testing and allozyme screening reveal that, among plants, forest tree species are generally highly heterogeneous, but the genetic variation is organized within and between populations in diverse ways.

Variation Within Populations

Information about genetic variation within populations is largely based on allozyme surveys, which examine forms of enzymes that can be distinguished by the technique of electrophoresis. Three measures are generally used to estimate variability: (1) expected heterozygosity in random breeding populations or the observed heterozygosity averaged over all loci, (2) proportion of polymorphic and monomorphic loci, and (3) the mean number of alleles per locus. In general, forest trees are highly heterozygous—the average level of heterozygosity is twice that in herbaceous plants (Hamrick et al., 1979). Similarly, the proportion of polymorphic loci (those for which numerous alleles exist) is also high; most species show allelic variation at more than 50 percent of the studied loci (Table 3-2).

Although genetic variation has been surveyed in a wide range of species, including conifers and some tropical and temperate angiosperms (flowering plants), the existing forestry data base is largely derived from the study of conifers in the north temperate zone (Ledig, 1986; Zobel and Talbert, 1986). Among the few exceptions to the general finding of high levels of genetic variation in trees are *Pinus resinosa* (red pine) (Fowler and Morris, 1977) and *Pinus torreyana* (Torrey pine) (Ledig and Conkle, 1983). Populations of *Abies bracteata* (Santa Lucia fir) also have low levels of genetic variation (Ledig, 1987).

All populations of a species do not have the same level of genetic variation. Peripheral populations in *Picea abies* (Norway spruce) (Bergman and Gregorius, 1979) and *Pinus contorta* (shore pine) (Yeh and Layton, 1979), for example, have been shown to contain less genetic variation than the central populations. This occurrence is not well documented in other species and may merely reflect smaller population sizes at the periphery. Peripheral populations are also usually under a more rigorous selective regime than their central counterparts. Genetic characteristics of such populations may, therefore, be important in understanding how populations may respond to altered selective pressures.

Finally, the level of genetic variation has also been found to be correlated with several life history and environmental parameters (Hamrick et al., 1979; Loveless and Hamrick, 1984). In general, widely spread plant species with large ranges that are outcrossing have high levels of genetic variation. Among the species with low levels of variation, *Pinus torreyana* and *Abies bracteata* have restricted ranges, but *Pinus resinosa* is widely distributed. Understanding the subtle factors that induce and maintain genetic diversity provides the scientific foundation for the optimal management of existing tree genetic variation.

Estimates of Population Genetic Structures

There is growing evidence that various genotypes in plant populations are not randomly distributed. Nonrandom distribution of genotypes in a population can make diverse sampling difficult. Selection and mating among related individuals (siblings or cross-generational relatives) in close proximity can affect genetic structure and thus sampling within the population.

For several species, there are more homozygous individuals than would be expected from random mating, even within plots of 1 ha or less. This cannot be explained by selfing or other forms of inbreeding. In one such species, *Pinus sylvestris*, the excess of homozygotes could be due to very localized and specific male effectiveness (Tigerstedt, 1984). This appears to be true for *Liriodendron tulipifera* (Brotschol, 1983) but only at the seedling stage, and it is not evident in the adult trees. Such effects appear to be even more pronounced in some tropical tree species, probably due to more limited pollen dispersal by animals (O'Malley and Bawa, 1987). Pollen flow in these species appears to be very extensive, but the pollen distribution of individual trees may be highly localized.

The net effect is that seed and seedling populations may be highly structured. The implication for sampling is that unless the sampling is spread over a wide area within the population, the collected seeds may represent a biased sample of genetic diversity within the population.

Variation Among Populations

Just as variation in species has implications for management, so does the distribution of that variation through the species. Forest tree species differ greatly with respect to genetic divergence among populations. There are two major problems in comparing levels of genetic differentiation within a species. First is the problem of scale; in some species, sampled populations are separated by several hundred kilometers, in others by only a few kilometers. Second, genetic divergence in some species has been studied on the basis of variation in morphological traits and in others by estimating variation in allozymes. The results obtained from the two approaches frequently do not agree.

In many species, the evidence for genetic divergence is largely based on allelic diversity as revealed by allozyme studies. Several methods

TABLE 3-2 Genetic Diversity in Tree Species

Species	Expected Heterozygosity[a]	Number of Loci Assayed	Percentage Polymorphic of Loci		Sample Size	Origin
			A[b]	B[c]		
Pinus (pines)						
P. attenuata	0.125	22	73	27	10 pops.	Rangewide
P. attenuata	0.087	43	58	49	10 pops.	Rangewide
P. banksiana	0.141	27	74	—	32 pops.	Ontario
P. banksiana	0.115	21	81	52	3 pops.	Alberta
P. brutia	0.130	29	45	38	10 pops.	Rangewide
P. caribaea	0.212	18	72	48	7 pops.	Central America
P. contorta	0.184	21	86	71	5 pops.	Alberta
P. contorta	0.160	25	59	45	9 pops.	British Columbia, Yukon
P. contorta[d]	0.185	39	90	44	1 pops.	California
P. contorta	0.116	42	68[e]	—	32 pops.	Rangewide
P. contorta	0.135	9	44	44	4 pops.	Within 2 km in Colorado
P. coulteri	0.148	33	49	45	8 pops.	Rangewide
P. halepensis	0.040	28	21	21	19 pops.	Rangewide
P. jeffreyi[d]	0.261	43	86	67	4 pops.	Central California
P. jeffreyi	0.255	20	90	90	14 pops.	No. California, Oregon
P. lambertiana	0.275	19	79	58[d]	58 individuals	California
P. longaeva	0.327	14	79	—	5 pops.	Nevada, Utah
P. monticola	0.180	12	65	51	28 pops.	Rangewide
P. muricata	0.084	46	67	56	18 pops.	No. California
P. nigra	0.272	4	100	75	28 pops.	Yugoslavia, Mediterranean
P. oocarpa	0.183	18	72	55	8 pops.	Central America
P. palustris	0.150	19	100	84	24 pops.	Rangewide
P. ponderosa	0.124	23	74	35	6 small, isolated pops.	Montana

P. ponderosa	0.186	29	90	32	400 trees	Washington, Idaho, Montana
P. ponderosa	0.123	21	62	38	10 pops., pooled	Washington, Idaho, Montana
P. radiata	0.126	37	70	51	3 pops.	California
P. resinosa	0.007	27	15	4	2 pops.	Minnesota
P. resinosa	0.002	46	—	0	50 trees	Wisconsin
P. rigida	0.146	21	100	76	11 pops.	Rangewide
P. sabiniana	0.128	29	93	62	8 pops.	Rangewide
P. strobus	0.236	12[f]	83	66	27 pops.	Rangewide (grown in common garden)
P. strobus	0.330	17	53	—	35 selected clones	—
P. sylvestris	0.309	16	100	81	14 pops.	Scotland
P. sylvestris	0.310[g]	11	100	91	9 pops.	Sweden
P. taeda	0.362	10	100	90	1 pop.	North Carolina
P. taeda	0.282	25	96	80	90 selected clones	Southeastern United States
P. torreyana	0.000	59	0	0	2 pops.	Rangewide
Picea (spruces)						
P. abies	0.220	34	82	65	9 pops.	Poland
P. abies	0.410[g]	7	—	—	21 pops.	Northern Europe
P. glauca	—	26	77	52	Several pops.	Alberta
P. glauca	0.140	20	—	—	1 pop.	Alberta
P. sitchensis	0.150	24	51	46	10 pops.	Oregon, Alaska
Abies (firs)						
A. alba	0.500[g]	9	100	94	4 pops.	Czechoslovakia
A. balsamea	—	14	57	—	4 pops. within 3 km	New Hampshire
A. balsamea, A. fraseri, and transitional pops.	0.130	20	65	60	12 pops.	Eastern United States
A. bracteata	0.052	30	37	23	5 pops.	Rangewide

(continued)

TABLE 3-2 (Continued)

Species	Expected Heterozygosity[a]	Number of Loci Assayed	Percentage Polymorphic of Loci		Sample Size	Origin
			A[b]	B[c]		
Other conifers						
Calocedrus decurrens	0.180	25	96	76	12 pops.	California
Cupressus macrocarpa	0.160	28	61	61	2 pops.	Rangewide
Larix decidua	0.081	28	50	36	11 pops.	Rangewide
Larix occidentalis	0.074	23	39	26	19 pops.	Washington, Idaho, Montana
Larix leptolepis	0.073	16	37	25	9 pops.	Rangewide
Pseudotsuga menziesii	0.331	17	100	88	1 pop.	Oregon
Pseudotsuga menziesii	0.155	21	86	67	11 pops.	British Columbia
Sequoiadendron giganteum	0.140	8	50	50	34 pops.	Rangewide
Thuja plicata	0.040	—	—	—	—	—
Eucalyptus (Australian eucalypts)						
E. caesia	0.120	18	61	44	13 pops.	Rangewide
E. cloeziana	0.240	—	—	—	—	—
E. delegatensis	0.270	—	—	—	—	—
E. grandis	0.180	—	—	—	—	—
E. saligna	0.260	—	—	—	—	—
Tropical angiosperms						
Acalypha diversifolia	0.273	29	68	—	3 pops.	Barro Colorado Island
Alseis blackiana	0.374	26	90	—	3 pops.	Barro Colorado Island
Hybanthus prunifolius	0.247	42	70	—	3 pops.	Barro Colorado Island
Psychotria horizontalis	0.152	20	50	—	3 pops.	Barro Colorado Island

Quadrartibea asterolepis	0.256	30	64	—		3 pops.	Barro Colorado Island
Rinorea sylvatica	0.106	35	35	—	—	3 pops.	Barro Colorado Island
Sorocea affinis	0.239	36	72	—	—	3 pops.	Barro Colorado Island
Swartzia simplex var. *ochnacea*	0.272	36	76	—	—	3 pops.	Barro Colorado Island

NOTE: Genetic diversity calculations were based on enzymes of the haploid female gametophyte for conifers and diploid leaf tissue for angiosperms, except as noted. The — symbol indicates that the information was unavailable; pops. = populations.

[a] Expected heterozygosity was calculated as the mean of within-population heterozygosities, except as noted.

[b] Proportion of loci polymorphic under the criterion that a locus was polymorphic if any allelic variant was observed.

[c] Proportion of loci polymorphic under the criterion that the most common allele was present in a frequency less than 0.95 in every population.

[d] Total heterozygosity and percentage polymorphic loci were calculated from allele frequencies averaged across populations.

[e] Proportion of loci polymorphic under the criterion that the most common allele was present in a frequency less than 0.99.

[f] Genic diversity calculations were based on foliar enzymes.

[g] Only variable loci were used.

SOURCE: Adapted from F. T. Ledig. 1986. Heterozygosity, heterosis, and fitness in outbreeding plants. Pp. 77–104 in Conservation Biology: The Science of Scarcity and Diversity, M. E. Soulé, ed. Sunderland, Mass.: Sinaur Associates. Reprinted with permission.

Knowledge about the distribution and reproductive biology of tropical forest trees is gained through botanical surveys and plant inventories in the tropics. Here a research assistant climbs a tree of the *Tabebuia* sp. to collect botanical specimens. Credit: Calvin R. Sperling.

exist not only for assessing the average levels of differentiation (Gregorius, 1984), but also for determining which loci or which populations are most deviant (Gregorius and Roberds, 1986). Among these, Nei's measures are most commonly reported (Nei, 1972).

It can be shown that species vary considerably in terms of both the patterns and the levels of genetic variability among populations. Within most species, some hierarchy of subdivisions exists due to interpopulational segregation within localized plots and to individual tree inbreeding, as in *Abies fraseri* (Ross, 1988). Typically, widespread species, such as *Pseudotsuga menziesii* (Douglas fir), show geographical or ecotypic differentiation as, for example, between coastal and inland populations. Moreover, within each geographical area, large interpopulational differences may exist. Such species may thus show high variability within and among populations. Other species, such as *Pinus longaeva* (Great Basin bristlecone pine), show high variability within populations but low differentiation among populations (Hiebert and Hamrick, 1983). In species such as *Pinus resinosa* and *Abies bracteata*, the level of variation is low both within and among populations. By contrast, the two populations of *Pinus torreyana*, each of which is genetically uniform, are quite distinct from each other.

The level of genetic differentiation may also vary within a species. *Pinus monticola* (western white pine), for example, is highly variable in California, but not in other parts of its range (Steinhoff et al., 1983).

A very preliminary analysis of a few tropical forest tree species reveals that levels of population genetic divergence are similar to those observed for temperate tree species. This is contrary to expectations based on the limited gene dispersal that might be typical of many tropical species whose populations are more isolated (Bawa, 1976).

The level of genetic differentiation among populations has been quantified in many species. In most species, particularly the widespread ones, 2 to 16 percent of the total genetic variability is due to interpopulation differences (Fins and Seeb, 1986; Moran and Hopper, 1987).

Mating systems and the geographical range of species seem to have significant effects on the level of genetic variation among populations (Hamrick, 1983). Self-pollinating species show more interpopulational divergence than outcrossing species (Hamrick, 1983). Regional and localized species also exhibit greater differentiation among populations than widespread species (Moran and Hopper, 1987).

The conservation implications of the available data on genetic variation are that, in the case of outcrossing and widespread species, a few populations in each geographical zone may conserve much of the genetic diversity. For inbred and regional or localized species, however, many populations may be required to conserve a significant proportion of genetic variation (see Chapter 4).

Allozyme Studies and Morphological Variation

It is not known whether genetic differentiation inferred from allozyme studies is neutral, as generally assumed, or if it is indicative of a selective advantage or disadvantage. This is important because if much of the variation seen in allozymes were to be devoid of evolutionary significance (neutral), it would be difficult to justify maintaining multiple populations of trees to preserve their potentially useful genetic diversity based solely on allozyme variations. Unfortunately, allozyme studies in forest trees generally have not been accompanied by parallel work on variation in morphological traits that are known to contribute to the fitness of plants. In some studies, agreement between variation in allozymes and variation in other quantitative traits has been reported (Hamrick, 1983). In several species, however, the level of genetic divergence revealed by variation in allozymes is not correlated with morphological variability detected by provenance research (Libby and Critchfield, 1987; Moran and Adams, 1989; Namkoong and Kang, 1990). In particular, large-scale genetic variation along a geographic gradient (clinal variation) in *Picea sitchensis* (Sitka pine) or the extensive variety of ecotypes in *Pseudotsuga menziesii* is not revealed by allozyme studies (Falkenhagen, 1985).

Decades of provenance research have shown that most forest tree species exhibit considerable population divergence in genetically based traits of direct survival value. There remains, however, a need to conduct parallel allozyme and morphometric studies to determine to what degree data on isoenzymes can be used as indicators of morphological variation, or to monitor changes in gene frequency due to environmental or ecological events.

Allele Frequencies, Gene Flow, and Selection

The importance of interpopulational differentiation lies not only in whether alleles would be missed by sampling only a few populations, but in the occurrence of different combinations of alleles and their frequencies in different populations. This could make certain alleles or combinations of alleles easier to obtain from some populations than others. For outcrossing species with both large population sizes and high migration rates, the existence of alleles unique to a population is unlikely. In such species, alleles are likely to be widely distributed

Long-term ecological studies on the dynamics of species loss in tropical forests are being conducted so that reserves of adequate size may be established to perpetuate species diversity. A joint project of the Brazilian Instituto Nacional de Pesquisas da Amazonia (INPA) and the U.S. World Wildlife Fund is studying reserves ranging from 1 ha (pictured above) to 10,000 ha in Brazil both before and after they are isolated by clearing adjacent land. Credit: Douglas Daly.

without much variation in average frequencies. Natural selection, however, can play a strong role in causing differentiation among populations, especially if the selection pressures are high, migration is low or episodic, or if some populations are disjunct and small.

The potential for wide seed or pollen dispersal affects genetic structure. For some species, even those as widespread as *Pinus taeda* in the southeastern United States, this may not always be true. Reproduction can, in some time periods, be episodic in relatively small patches and thus limit the population sample in each patch. This would generate a genetic and demographic mosaic. Such mosaics can disappear if widescale clearing occurs (Roberds and Conkle, 1984). Genotypic distribution of any reestablished population then depends on the mating pool extant within the limited time period of forest regeneration. For species with more restricted pollen or seed migration, mosaics within forest stands are expressed as large interpopulational variations of longer duration. Selection forces operating on differing gene frequencies can affect the distribution of genetic variation among and within those populations.

The maintenance of the heterogeneity of allele frequencies that results from selective differences in the absence of mating barriers is possible but usually very difficult. Thus, some restrictions on gene flow are usually necessary for populations to differentiate under natural conditions. Over large distances of dense stands, the area of interbreeding neighborhoods can be small for some tree species (Wright, 1962). In fact, most tree species do display some degree of interpopulational differentiation in at least some phenotypic traits. For breeding purposes, it may be of direct importance to capture and make use of such interpopulational differences.

The genetic structure of populations thus provides a good average view of the amount of subdivision among population samples. Mating structures can be complex, and no single statistic can explain all of the important features of allelic distributions. Often differences between pollen and seed migration rates exist, and the extent to which they coincide with selective differences affects the distribution of alleles, allelic extinction rates (rate of gene loss), and the probability distributions of unique alleles (Gregorius and Namkoong, 1983; Namkoong and Gregorius, 1985).

Differences also exist in the patterns of population subdivision indicated by isozyme versus morphological traits, in the traits associated with broad climatic adaptability (e.g., bud break or bud set times), and in the traits associated with local site differences, such as elevation. The boundaries of selection do not act consistently among traits or among loci. There is often reason to suspect that local, naturally occurring populations are neither optimally reproductively fit (Eriksson et al., 1972) nor ideal for selection in production forestry breeding, even within their area of origin (Namkoong, 1969). Natural occurrence of a species in a region, therefore, should not be the sole criterion for its development as a production crop.

Population Complexity

Populations of trees evolve from different genetic structures and with the confounding effects of inbreeding and selection. A major analytical difficulty is to discern which factors are most important in generating the current and evolving patterns of genetic distribution in the population. Given the complexity of the mating systems of many tree species and the ecological instability of stand boundaries and gap distributions over time, great potential exists for many locally differentiated populations to have existed in the past or to be generated in the future. This can significantly complicate sampling the current variation. It must also be considered when managing a sampled set of genotypes or conservation stands.

CONCLUSIONS

Tree species are most commonly and effectively managed in planted stands. In many instances, trees will be maintained in or near their natural habitat and be classified as in situ. They can, for a number of objectives, be grown and sustained in new environments ex situ. For both types of management, it is of fundamental importance to understand the biological factors that influence the structure of genetic variation within tree populations. The current state of knowledge about breeding systems is sparse, as is the knowledge of the structure of populations, especially in tropical species. In nature, trees exhibit a high degree of heterogeneity (genetic differences), and the goal of conservation and management activities must be to capture and maintain that nonuniform genetic configuration.

The criteria to be used for sampling trees to be conserved and maintained will depend on the breeding system of each species. If knowledge of the reproductive biology of a species is incomplete, collection activities should be conducted at all extremes of population occurrence and include a high level of redundancy, even in the central areas of occurrence. The designation of large areas for in situ conservation is especially critical in tropical regions where tree species occur in lower densities as part of complex ecosystems that are biologically poorly understood.

RECOMMENDATIONS

To support conservation efforts, study of the patterns of genetic variation in tree populations should be accelerated and expanded in scope, especially in the tropics and subtropics.

Although genetic variation has been surveyed in a wide range of species, the existing data base is largely derived from the study of conifers in the north temperate zone. Similar data for tropical regions are lacking. A very preliminary analysis of a few tropical forest tree species reveals that levels of genetic divergence within populations are similar to those observed for temperate tree species. This is contrary to what the very different reproductive mechanisms of many tropical species have suggested. Thus, further study is needed before this information can be generally applied to conserving tropical tree species.

Research is needed to elucidate the distribution and structure of genetic variation—especially for tropical trees—and to support conservation efforts. Study of the patterns of genetic variation in tree populations should be accelerated and expanded in scope, especially in the tropics and subtropics. Genetic management and conservation programs should be applied to many more species and populations, particularly those that are endangered and those that are potentially useful. Inventories of genetic resources and patterns of variation should be accelerated and expanded to more areas, especially in the tropics. Genetic management techniques for resource areas that are managed for agroforestry and for industrial forestry should be studied to develop inexpensive means for monitoring the distribution of genetic variation and ensuring the maintenance of genetic diversity.

Research on population sizes, structures, dynamics, and reproductive systems should also be expanded, particularly in the following three areas:

• Research should be increased on reproductive systems, the genetic architecture of populations, and the minimum viable population size of trees. Such information is essential for the design and management of in situ and ex situ conservation stands.

• Knowledge must be gained of key interactions in complex tropical communities to support management of in situ stands and to prevent large-scale changes in population structures and species composition.

• Efforts should be made to anticipate the effect of global climatic change and pollution on the geographical distribution of species and on in situ reserves.

4

Conservation and Management of Tree Genetic Resources

nowledge of the structure of genetic variation in species is needed to make decisions about how to protect the genetic diversity of trees. Without that knowledge, the safest conservation strategy requires conserving virtually everything, without any priorities. Some aspects of the distribution of genetic variation are known, however, and that information, while incomplete, can help guide the development of useful conservation strategies.

CONSERVATION AND MANAGEMENT STRATEGIES

The structure of genetic variation within and among species is an inherent feature of the evolution of forests and must therefore be considered in developing any conservation strategy. It is also directly and indirectly influenced by many human activities, from incidental and unintended effects to intensive management and breeding. Breeding can create greater diversity among populations and can enhance the utility of the genetic resource by managing advanced generations of diverse breeding populations. In contrast to agricultural crops, forest trees have long regeneration cycles and generally grow under less intensive field cultivation. Their breeding, therefore, requires greater use of wide sources of variability. Less intensive activities than breeding can still conserve the present distribution of genetic variation, or can guide the future evolution of at least some parts of forest ecosystems by affecting natural regeneration.

When conserving trees in situ, it may be necessary to incorporate very large areas of land in order to conserve the gene pool adequately, because of the potentially wide geographic distribution of diversity and the complexity of the mating systems involved. Merely counting numbers of trunks is an inadequate guide to determining the effective population sizes of species, and merely counting species is inadequate for determining the existence of the genetic resources of key plant and animal species. Knowledge of the requirements for perpetuating most tree species is generally meager, and ability to organize conservation programs is low. As a result, conservation is often reduced to preserving areas in centers of species diversity in the hope that genetic diversity and differentiation are also conserved. Clearly, this pragmatic approach is necessary, but by itself, it is an insufficient step in genetic conservation and only adequate as a short-term contingency measure while more satisfactory conservation strategies are developed.

A few species, such as *Picea abies* (Norway spruce), are already included in broadly sampled collections. Others, such as *Pinus densiflora*, have such little variation among populations that there is little current concern for conserving their genetic variation. For most species, however, genetic variation is not well conserved, and certainly, not all significant variation is included in established genetic conservation or breeding programs. Further, for scores of species, ecogeographical surveys are still needed, at least to target populations for breeding. Such surveys are also needed for continued monitoring of the distribution of genetic variation to trigger warnings about the need for management interventions. If the sampling problem is solvable, then an array of management techniques exist that differ not only in the details of their execution but also in the conditions under which they are necessary or particularly useful.

At one extreme of simplicity, a species may exist in a state of homogeneity for all gene loci so that any sample of sufficient size would capture all alleles. In such situations the breeding can be straightforward, based on a single selection objective. Such a situation may exist for a few species whose natural variation may be very low. Also, for genetically depauperate species, that is, species in which little useful genetic variation exists, any equivalent size sample is as sufficient as any other sample. In such cases, simple storage and propagation programs are sufficient.

For almost all tree species, however, experimentation shows that genetic variation is high. But not all species have to be developed for all uses; hence, if the objectives are limited, finite conservation programs can be more readily defined. Some species exist within secure collections that contain a wide sample of the extant genetic variation and little further collection is needed, although better maintenance may be required. If species are classified according to the objectives of a forest management program, those whose values lie exclusively in nonproduction functions can only be managed in situ and are primarily reproduced by natural regeneration. For most of these species, no direct management interventions are feasible, but some forms of forest management (by regulating removal of trees or by preventive maintenance) can affect population sizes and structures, as well as their genotypic distributions, and thereby maintain the genetic variation needed for population viability and general evolution of the species. It is also possible that introducing populations of key species into reclamation areas or stand clearings could affect the evolution of the local ecosystem.

For most of the tens of thousands of species whose values are unknown, conserving genetic variation also depends on maintaining in situ stands. The adequacy of such programs for conserving nonproduction functions would therefore be the primary focus for these species. Although some of these species might eventually be found to be amenable to production forestry and some will be found to have traits of use for timber, medicinal, or other products, their interim maintenance will largely depend on the quality of in situ programs. Seed storage may be a feasible means of conserving sampled variation, and it may be necessary for species in endangered habitats. The cost of sampling, collecting, and storing more than a few hundred or thousand species may be too high, however, to justify allocating scarce funds for this purpose. Moreover, storage methods are not known for seeds of most species.

IN SITU CONSERVATION

In general, in situ conservation methods share three characteristics (Food and Agriculture Organization, 1984a):

• All growth phases of a target species are maintained largely within the ecosystem in which they originally evolved.

• Land use of the sites (e.g., agroforestry) is limited to those activities that will not have detrimental effects on habitat conservation objectives.

• Regeneration of target species occurs without human manipulation, or intervention is confined to short-term measures to counter detrimental factors resulting from adjacent land use or from fragmentation of the forest. Examples of manipulation that may be necessary in heavily altered ecosystems are artificial regeneration using local seed and manual weeding or controlled burning to suppress competing species temporarily.

WHY TROPICAL TREE SPECIES MAY NEED LARGER RESERVES

Protium occultum *Daly is a rare tree known only from Manaus and Jari in Brazil. It was first collected in 1985. The species belongs to the Burseraceae (frankincense) family, which throughout the tropics provides sources of resins used for illumination, caulking agents, and medicines. Credit: Douglas Daly.*

Ample evidence exists that canopy trees in most forests, including tropical lowland wet forests, are strongly outbred (Ashton, 1969; Bawa, 1974, 1979; Bawa et al., 1985). However, population densities of large trees, which constitute the most important forest genetic resources, in tropical lowland wet forests are extremely low. In a tropical lowland wet forest in Panama, for example, Hubbell and Foster (1986) found that one-third of all plant species with individuals larger than 1 cm dbh (diameter at breast height) were represented by only one adult per ha in the 50-ha plot they sampled. Assuming that these species are evenly distributed (many are not), an area of 20 km^2 would be required to encompass 2,000 individuals. For the rarest tree species in southeast Asian tropical wet forests, Ashton (1981) estimated that an area of 20 km^2 would be required to encompass 200 individuals. By extrapolation, 200 km^2 may be needed for 2,000 individuals.

Unfortunately, in many regions the areas earmarked for conserving certain vegetative types or forest genetic resources are well below the minimum sizes estimated above for a single species in contiguous areas. Most species are not uniformly distributed, and temporal variation within any possible reserve area may also exist. Therefore, it is logical to consider the minimum area required to maintain genetically viable populations in one or several reserves. Moreover, rare species may not, in fact, be rare everywhere. Thus multiple reserves may conserve more rare species than single reserves. Even so, the reserves would have to be very large for some species.

Key requirements for in situ conservation of threatened or endangered genetic resources are the estimation and design of minimum viable population areas for a target species. To ensure conservation of substantial genetic diversity within a species, multiple reserves must be created, the exact number and size of which will depend on the distribution of the genetic diversity of the selected species. The promotion of the continued maintenance and function of an ecosystem under in situ conservation depends on an understanding of several ecological interactions, particularly the symbiotic relationships among plants, pollinators, seed dispersers, fungi associated with tree roots, and the animals that live in the ecosystem.

Minimum Viable Population Size

The concept of minimum viable population size implies that a population in a given habitat cannot persist if the number of organisms is reduced below a certain threshold. It is a complex concept because there is no recognized minimum viable population size for most species. Whether a population of a given size can persist depends on a number of random or unpredictable demographic, genetic, and environmental events (Gilpin and Soulé, 1986). Moreover, population size varies with such attributes as life history, particularly generation time and the breeding system, and the spatial distribution of resources (Gilpin and Soulé, 1986). Nevertheless, minimum viable population sizes have been estimated for several groups of organisms on the basis of genetic criteria (Franklin, 1980; Soulé, 1980).

Three broad approaches to estimating minimum viable population size have been taken. One approach is to estimate the effective population on the basis of ability to withstand loss of genetic variability due to small population size. For animal populations, it has been estimated that loss of genetic variability due to inbreeding can be avoided if the rate of inbreeding per generation, F, is kept below 2 percent (Franklin, 1980; Soulé, 1980). If F is known, the effective population size (N_e) can be calculated by the equation:

$$F = \frac{1}{2N_e}$$

Thus, the effective population size of 25 would be sufficient if 2 percent inbreeding per generation is acceptable. Taking 1 percent as a more conservative estimate of a tolerable level of inbreeding in animals, Frankel and Soulé (1981) calculated the minimum population size to be 50. This effective population size is in general sufficient for short periods (i.e., a few generations), after which the captive populations can be released in the wild and variation might increase. However, the applicability of this approach as well as the estimated effective population size to forest trees is questionable. Such mathematical approaches can oversimplify more complex biological realities (Ewens et al., 1987). Although the population sizes are on the same order of magnitude as those derived from ecological models, the effects of demographic randomness on the total sizes needed are larger due to the independent factors of inbreeding and random loss in the population.

The second approach is to estimate effective population size on the basis of the number required to maintain the evolutionary potential of the population. It has been estimated that if N_e is 500 individuals, a panmictic population—one in which mating is entirely random—is not likely to lose genetic variance due to drift and can retain enough variation to respond to altered selection pressures (Franklin, 1980). Assuming that the ratio between census numbers N and N_e is 3 or 4 (Soulé, 1980), the minimum population size then becomes 1,500 or 2,000 individuals. Both the N and N_e specified above are for outbreeding, monoecious species.

The third approach is based on calculating the population size that will minimize the sampling loss of alleles that occur in low frequency. Namkoong (1984) has estimated that in species with known levels of inbreeding and population structure, a sample size of 1,000 will keep the probability of the loss of an allele that occurs at the frequency of 0.01 at a particular locus below 0.01. An increase in the number of loci at which desired rare alleles occur will increase the number of individuals required to minimize the probability of loss, but at a much lower rate than by decreasing the frequency of the desired allele.

The sizes mentioned above are based on the minimum population sizes required for evolutionary flexibility and, therefore, continued survival. However, minimum population size is a probabilistic concept and not a fixed number, and can be affected by biological, environmental, or genetic events (Soulé, 1987). It specifies a population's probability of becoming extinct under certain conditions. That probability depends not only on the past evolutionary history of the species and its current genetic structure, but also on demographic and environmental randomness (Gilpin and Soulé, 1986). The minimum population size, therefore, is likely to differ among species and among habitats for the same species.

Number of Reserves and Sampling Strategy

Although much of the literature is couched in terms of conserving particular populations, in situ conservation in reality involves preserving whole communities. The number of populations and species that require some protective measure in the wild is so large that it is impractical to design in situ conservation programs on the basis of individual species and their populations. There may exist well-correlated sets of co-occurrences of species that can, for immediate conservation purposes, be considered to be distinct assemblages, if not communities.

In areas where several species are being simultaneously conserved in a reserve, a problem exists in ensuring that the number and distribution of such populations of the contained species are adequate for maintaining genetic diversity in either single or multiple reserves. Without information about the distribution of genetic variability, it is difficult to assess the number and distribution of populations in one or more reserves that might be required to encompass much of the genetic diversity. This lack of information also constrains determining the level of migration for various species. For some species this may be close to zero while for others it may be high. The capacity of individual reserves to preserve evolutionary dynamics within their individual borders can have significant effects on migration.

Because forest trees generally show interpopulational variation in some traits, several small reserves spread over a large geographical area may conserve total genetic diversity more effectively than a single, large reserve. Theoretically, and for easily managed species, the viability of single populations may be maintained with effective population sizes of 50 to 100 reproductive individuals, and it would be possible to contain a total of a few thousand in as few as 50 or so reserves. For many tropical tree species, however, areas with 50 to 100 individuals may be too small to maintain the integrity of key mutualistic interactions and to preclude instability due to random events. Larger populations must then be maintained, or means must be developed to augment seed and pollen migration, such as planting or providing corridors for seed and pollen dispersers. Thus, a reserve system that is adequate for any one species may be insufficient for others.

To determine the adequacy of a reserve system, an inventory strategy can be used. A finite number of species clusters, cover types, or key species could be inventoried and mapped, and the areas required for adequate allele sampling could then be designated. Simultaneously, an inventory of existing reserves and parks could be made to determine the extent to which populations are included within those same cluster categories. A comparison of the sets would then indicate the extent to which populations of those indicator clusters are already included in reserve areas and what other reserve areas are needed for the sampled species. An extrapolation to all species could then be estimated on the basis of the representativeness of the original sample. This approach lacks the population- and allele-level detail needed for all species in a complete program, but it could be used to identify the magnitude and location of major genetic deficiencies at the population level.

Segregating Populations

Within reserves, populations can be identified in patches of varying sizes. With tree species, both temporal and spatial variations can affect

---

CATEGORIES FOR CONSERVATION AREAS

The International Union for the Conservation of Nature and Natural Resources (IUCN) classifies in situ conservation areas on the basis of 10 management categories (1978): I, scientific reserve/strict nature reserve; II, national park; III, natural monument / natural landmark; IV, nature conservation reserve/ managed nature reserve/wildlife sanctuary; V, protected landscape or seascape; VI, resource reserve; VII, natural biotic area/anthropological reserve; VIII, multiple-use management area/managed resource area; IX, biosphere reserve; and X, world heritage site (natural). The management methods for four of those categories (I, II, IV, and VIII) are pertinent to this discussion and IUCN's descriptions are presented below.

Among national parks in Canada, Banff National Park is the oldest. It was established in 1885. Credit: George F. Mobley ©National Geographic Society.

Strict Nature Reserves

The objectives of a strict nature reserve are to protect communities and species and to maintain natural processes in an undisturbed state in order to have ecologically representative examples of the natural environment available for maintaining genetic resources in a dynamic and evolutionary state and for scientific study, environmental monitoring, and education. Such reserves often contain fragile ecosystems or life forms, are areas of genetic structure, but they can be managed to differentiate population segments. To maintain genetic diversity in a single reserve, sufficient numbers of interbreeding but segregated (separate) populations are needed. The above-stated figures on minimum effective population sizes would then have to be inflated if populations are endangered by common threats. Segregated populations may also be required to generate wider variations and to protect species-wide allelic diversity more effectively. Hence, reserves should be large enough or managed

---

important biological or geological diversity, or are of particular importance to conserving genetic resources. Because only natural processes are allowed to take place, without any direct human interference, only monitoring and nondisruptive sampling are permitted. Because extinction is a natural process, for species that are vulnerable such reserves are a resource only as a supplement to other conservation programs.

National Parks

The management objectives of these types of areas call for protecting natural and scenic areas of national or international significance for scientific, educational, and recreational uses. To ensure ecological stability and diversity, the areas should perpetuate, in a natural state, representative samples of geographic regions, biotic communities and genetic resources, and species in danger of extinction. Such areas generally encompass relatively large land tracts that contain one or several entire ecosystems that are not materially altered by human exploitation and occupation. As for managed nature reserves with secure legal status, opportunities exist for developing minimally interventionist techniques for ensuring genetic diversity in self-sustaining ecosystems. Possibilities also exist for developing genetic diversity in restored ecosystems by introducing certain levels and types of genetic variation. With sustained monitoring, these areas can also serve as a secure collection of evolving populations.

Managed Nature Reserves

The purpose of managed nature reserves is to ensure through specific human manipulation the perpetuation of natural conditions necessary to protect nationally significant species, groups of species, biotic communities, or physical features of the environment. Scientific research, environmental monitoring, and educational use are the primary activities associated with this category. Although a variety of protected areas fall within this category, each would have as its primary purpose the protection of nature and not the production of harvestable, renewable resources, although the latter might be an aspect of managing a particular area.

(continued)

in separate forest compartments to permit independent population development. If selection affects allele distributions, separate populations probably offer greater protection for allelic polymorphisms than do unified populations.

The design of conservation programs is still very primitive with respect to ensuring the structural integrity of the genetic system of conserved species in other than the simplest boreal types of ecosystems, if even there. A large degree of genetic redundancy within and among

---

Managed nature reserves can provide long-term security for species by maintaining the ecosystem necessary for natural reproduction or by introducing genotypes from other sources or from populations bred for different objectives. The development of genetically variable populations in seminatural (i.e., managed) ecosystems is a unique but as yet unexplored technique for simultaneously affecting and studying the responses of ecosystems to nonintentional interactions. These areas are often legally secured and, hence, can serve as a long-term, living resource base for the conservation of evolving populations.

MULTIPLE-USE MANAGEMENT AREA/
MANAGED RESOURCE AREA

This management category primarily supports economic activities, although specific zones may also be designated within an area to achieve specific conservation objectives. Parts of an area may be settled and may have been altered by humans. The goal of this management method is to provide for the sustained development of water, timber, wildlife, pasture, and outdoor recreation and at the same time provide for economic, social, and cultural needs over a long term. The areas are managed on a sustained-yield basis. Hence, they are often closely allied with traditional breeding operations for managing genetic diversity. In contrast to industrial forestry, the objectives of forest management and genetic selection in these areas are not exclusively for industrial profit. The managed populations would at least partially be recurrently selected and could be a diverse living resource.

Several programs, such as the United Nations Educational, Scientific, and Cultural Organization's Man and the Biosphere Program, with its "biosphere reserves," incorporate various mixtures of in situ techniques with the objective of ensuring the continued existence of well-protected areas and the continued use of resources by locally affected people. Often, the strict nature reserves are located in a protected zone, and surrounding or adjacent areas are designated for increasing levels of human intervention. Such systems can integrate multiple levels of genetic variation between and within usage zones and could provide unique opportunities for experimental genetic management.

conservation units is needed, at least until more is known about safe levels of loss and how they are balanced by changes in other units.

Knowledge of Ecosystem Dynamics and Integrity

A key requirement in managing nature reserves is the knowledge of ecosystem dynamics. The minimum viable population size may ensure genetic diversity through generations only if the structure and stability of the ecosystem are maintained. This is true regardless of the number of species that are targeted for conservation. Among the many biotic forces that impinge on community structure and stability is the integrity of the food web in an ecosystem (Pimm, 1986). The organization of food webs is particularly important for maintaining forest genetic resources in the subtropics or tropics, where the pollen and seed of an over-whelming majority of forest trees are dispersed by a wide variety of animals. The diverse feeding relationships among a multitude of animals and numerous plant species add extraordinary complexity to the web of dependency.

Species that provide food resources in the form of pollen, nectar, fruits, and seeds to pollinators, seed dispersal agents, and seed predators may play a critical role in maintaining the structure and stability of the community (Gilbert, 1980; Terborgh, 1986). In some areas, pollinators may come from distant sources and seed dispersers may migrate from one place to another on an elevation gradient. During the dry season in Costa Rica, for example, pollinating moths migrate from a dry deciduous forest to a wet forest several kilometers away (Janzen, 1987). In southeast Asian forests, bats fly over many kilometers to pollinate their host plants (Marshall, 1983). In Amazonia, fruit-eating species (frugivores) may migrate to other elevations during periods of food shortage (Terborgh, 1986). The populations of seed predators and seed dispersal agents may be regulated by top carnivores, which require a very large area to maintain viable populations. In reserves lacking carnivores, populations of seed predators and seed dispersal agents may increase with concomitant, but unknown, effects on the relative densities of various tree species (Terborgh, 1988).

In brief, maintaining tree populations in a tropical community is contingent on a very thorough understanding of key ecological inter-actions between plants and their pollinators, seed dispersers, and seed predators as well as the spatial and temporal distribution of floral, fruit, and seed resources. The link between pollinators, seed dispersers, and seed predators and their host plants is not the only critical element in maintaining community stability, although pollen-seed vectors and seed predators play a vital part in plant reproduction. Information on other processes and their effects on community integrity are contained in a series of articles in *Conservation Biology: The Science of Scarcity and Diversity* (Soulé, 1986).

Prospects for Effective In Situ Programs

An effective in situ program for genetic conservation has several key aspects: identification of gene pools, selection of specific sites, acquisition and design of the layout and administration of the reserves to ensure the availability of germplasm for use, and management of the reserves in perpetuity. Consideration of the needs of local people must also be emphasized, because in situ conservation techniques can rarely if ever be carried out without the collaboration of the locally affected people. Thus, adjunct programs that provide instruction in the purpose and use of conservation areas are often indispensable, and research is needed on the design of programs and social support structures that can contribute to maintaining in situ stands.

Because in situ stands often provide services to various interests, coordination among those interests must also be included in program development plans. This will often require that site selection criteria include ease of management and minimal disruption of local uses. For managed areas, sites must be integrable with multiple local uses and accessible for management and collection activities (Palmberg, 1988). For many tree species, for example, controlled harvesting or other moderate disturbances are not necessarily threats to viability.

The possibilities for translating these concepts into practice are strongly constrained by the diversity of funding sources for conservation activities and by the generally low levels of funds available. Moreover, many agencies, whether their programs are coordinated or not, have different conservation objectives. Hence, no global agenda for species inclusiveness exists. Only one-third of the biosphere reserves have been inventoried, even for trees; many other designated areas are not sufficiently well protected to ensure the security of species thought to be contained within their borders, and almost none is scientifically managed to maintain species diversity or genetic variation of the constituent species. In addition, no broad agreement exists on what is expected of local peoples or governments with respect to the conservation agency. Hence, the agency may confront lack of infrastructure support, indifference, or even an adversary relationship with different segments of the public and among sectoral users. The potential conservation utility of these programs has not been realized and may not be for many years.

Botanical gardens and arboreta are ex situ conservation sites for local and introduced tree species. The Arnold Arboretum in Jamaica Plain, Massachusetts, contains several species of the genus *Larix*, or larch (pictured here), and other species collected during plant exploration expeditions to temperate regions of Asia in the 1800s. Credit: Calvin R. Sperling.

EX SITU CONSERVATION

In contrast to in situ methods, ex situ methods include any of those practices that conserve genetic material outside the natural distribution of the parent population, and they may use reproductive material of individuals or stands located beyond the site of the parent population. Ex situ methods and materials include gene banks for seed or pollen and clonal banks, arboreta, and breeding populations (Bonner, 1985). They also include active collections involving shorter-term, temporary storage to distribute materials for evaluation and screening, as well as working collections for breeding.

The most common form of ex situ conservation of trees is the living stand. Such stands are frequently started from a single-source seed collection and are maintained for observational purposes. The size of the stands may range from specimens in botanical gardens and arboreta, to a few ornamental trees on small plots, to larger units with scores of trees.

Seed storage, another ex situ conservation method, refers to storage of intact seeds in a controlled environment. Under controlled temperature and moisture conditions, stored seed of some species remain viable for decades. This technique is the mainstay of germplasm conservation of agriculturally important species, and it is starting to be used for conserving rare tree species. When the viability of stored seeds decreases, the usual procedure is to regenerate the sample in the field. This procedure is impractical for tree species, however, because of the long vegetative period before trees produce seeds (up to 20 or more years), but alternative strategies can be developed. A system of regeneration stands can be organized, for example, to ensure the continuous availability of sexually mature samples, and vegetative materials can be held in juvenile condition by hedging and other techniques and then allowed to mature quickly when needed. Coordinated sets of materials can then be kept available for any set of stored genotypes.

Although collections exist in many independent tree seed banks, the collections are vulnerable because standards of maintenance are often less than ideal and regeneration is lacking or insufficient. Only the seeds with fully orthodox behavior, such as the seed of temperate-zone commercially important genera, store well for a few decades at subfreezing temperatures and 6 to 10 percent relative humidity. These include *Pinus* (pine), *Picea* (spruce), *Larix* (larch), *Abies* (fir), *Tsuga* (hemlock), *Pseudotsuga* (Douglas fir), *Alnus* (alder), *Fraxinus* (ash), *Liriodendron* (tulip tree), *Platanus* (plane tree), *Liquidambar* (sweet gum), *Betula* (birch), and *Prunus* (stone fruits). In the tropics, they include *Eucalyptus, Casuarina* (she oaks), *Citrus* (citrus fruit), *Gmelina,* and some *Dipterocarpus* and *Acacia* species. Some species produce recalcitrant seeds that are desiccation intolerant (cannot survive the removal of moisture), such as *Shorea* (mahogany) and some species within the genera *Quercus* (oak), *Acer* (maple), and *Aesculus* (horse chestnut) (P. S. Ashton, Harvard University, personal communication, May 1990; Bonner, 1990), or that cannot survive low temperatures, such as *Quercus* and *Aesculus* in the temperate zone and *Shorea* (mahogany) and *Hopea* species in the tropics

(Withers and Williams, 1982). Techniques such as cryopreservation of embryos, pollen, and tissue may enable long-term storage of desiccation-intolerant species.

Current seed collections are primarily stored for short- or medium-term storage of materials for afforestation and reforestation. Very few programs have long-term objectives, and at this writing, forestry programs have just begun to address the practical constraints of long-term seed storage.

With modern freeze-drying techniques, pollen of some species can be stored at a very low moisture content and at subfreezing temperatures. For regeneration purposes, however, this technique requires complementary female structures to enable use of the pollen in seed production. Strategies for the use of seed and pollen in regeneration still need to be defined and implemented. Although pollen seems harder to store than seed for some gymnosperms (e.g., conifers), the constraint may be overcome by further testing and development.

Maintenance of living stands as field gene banks is another ex situ method, but its use is currently restricted to highly selected genotypes of species that are of commercial importance, such as those in breeding programs in which graftings or rooted cuttings can be developed into mature reproductive materials. Hence, at present, such trees are hardly to be considered conservation collections. With the use of cuttings, genetic variation in cloning ability is often seen. Cuttings are useful for preserving specific genotypes, obtaining rapid regeneration, and saving genotypes faced with destruction that cannot otherwise reproduce.

Tissue culture also has potential to provide a secure conservation method. The technique involves micropropagation (whether meristems, embryos, or other). It requires large investments in development, but if cryogenic storage is developed it provides a secure conservation method. The concept of in vitro gene banks is being tested for selected agricultural crops by international agricultural research centers, and some operational aspects could be widely applicable particularly for facilitating the international exchange of disease-free planting materials (Withers, 1989). However, tissue culture is still in the experimental stages for most forest tree species.

Cryogenic storage, the preservation of biological material suspended above or in liquid nitrogen at temperatures from $-150°C$ to $-196°C$, has been used for many years as a means of keeping animal semen for breeding purposes. This technology is relatively new to seed storage, and hence, the time limits for storage of true orthodox seeds have not yet been determined. Cryogenic storage of genera with small seeds, such as *Eucalyptus*, may be cost-effective, and the technology promises

Biotechnology provides tools for managing forest tree resources. For example, genetically uniform plants can be generated in test tubes. This clump of pine tree plantlets was regenerated from individual plant cells in tissue culture. Credit: U.S. Agency for International Development.

less genetic damage than conventional seed storage. However, storage of recalcitrant seeds, costs for larger-seeded species, susceptibilities to mechanical breakdowns, and regeneration remain formidable problems to be overcome.

CHOICE OF METHOD

The use of ex situ stands for active conservation in multiple and well-secured locations could be applied to many more than the few species used thus far. Such stands could comprise relatively small areas, but they must be a part of a network of areas to ensure the survival and availability of propagules and to provide data on performance over a variety of sites. The design of such conservation stands—as active collections—is well known, but combining material from such collections with the working collections used in breeding programs and also with various managed areas used in in situ programs would provide a vital link for conserving and using the total gene pool of a species. While this is the underlying principle behind the networks of programs coordinated by the International Board for Plant Genetic Resources (IBPGR) for agricultural crops, there are no global programs for establishing such linkages for forest tree genetic resources.

One of the decisions that must be made now is to establish an active forestry program that creates an overall conceptual framework, including much-needed research on several of the above-mentioned technological barriers to safe, long-term seed storage. The knowledge of international organizations, such as the Food and Agriculture Organization (FAO), the International Union for the Conservation of Nature and Natural Resources (IUCN), the IBPGR, and many others, can be of great value in this regard. Forestry scientists will also be needed to convey information about the unique characteristics of forest species to agricultural scientists in the most efficient manner.

Technical and Biological Factors

Among the considerations that affect the choice of conservation technique is the fact that the technology does not exist for ex situ storage of many species, and thus, in situ management is necessary until such techniques have been tested and applied. On the other hand, growing trees to sexual maturity takes considerable time and space, as noted earlier, and regenerating populations with an intended mating and genetic structure can rarely be assured. Finally, although viable seed storage is not yet technically or economically feasible for many species, it is feasible for many other species, but it is not yet being implemented.

Effective use of conserved populations has exclusively depended on distributing seeds for testing and observation or as genetic material for breeding. A substantial time lag occurs before seeds can be grown into trees useful in breeding or other programs. This situation is also true

A Chinese forester explains the results of field growth performance tests of loblolly pine seed from the United States. Loblolly and slash pines are being planted extensively in Africa, Asia, and Latin America. Credit: Stanley Krugman.

when large initial seed collections or mature conservation stands are needed. Because of the time lag, needs for seeds or mature trees usually cannot be met quickly by growing out a population on demand, as is done with annuals. (However, some objectives could be achieved by wider use of vegetative propagation techniques, which could shorten the time lag.) Living stand collections have, therefore, been the technique of choice for short- to medium-term storage needs. For the long term (more than 50 years), seed storage is often preferred, but it requires the support of living stands. Long-term conservation by seed or tissue storage is not technically feasible for many species and, for those for which it seems promising, no formal program exists.

Among the biological factors that affect the choice of technique is the importance of natural mating and selection in the genetic structure of managed populations. For many species, the reproductive processes are difficult to control and yet strongly affect population structure. Selection forces resulting from multiple environmental stresses, competitors, mutualists, and pathogens may require the evolution of multiple in situ populations. With some conservation methods, it may be particularly desirable to allow the population to adapt to a relatively unmanaged ecosystem.

In situ conservation is also necessary to conserve communities and ecosystems, and it can be sufficient also for conserving the many plant and animal species associated with those systems. For research purposes and for use in uncontrolled environments, in situ populations are obviously needed. However, when breeding and selection systems are simpler and populations can be managed in controlled environments outside their ecosystems of origin, then breeding, testing, and germ-plasm development can be more efficient in ex situ populations that are developed for controlled-use conditions.

Management Factors

Among the management factors that affect the choice of technique is the capacity to protect or develop the resource. Given the uncertainty about the extent and distribution of genetic variation and, hence, the capability to target genetic sampling very well, in situ methods may require so many and such large areas that they exceed any reasonable management capacity. On the other hand, ex situ methods are limited by the capacity to store seeds, pollen, or tissue cultures and by managerial capacities to ensure survival and reproduction in controlled plantings.

Any in situ or ex situ methods that require large investments in area or effort for long, sustained periods are obviously vulnerable to lapses in control. The susceptibility of in situ methods to gene erosion seems most acute when habitats are threatened, ecosystems are unstable, and managerial authority is weak. Obviously, a great expansion of programs is needed in terms of both structural depth and species inclusiveness, but research on program design is also needed to estimate the effectiveness of in situ conservation. Ex situ methods seem of most limited value with species that are still maintained in wild or semidomesticated conditions and where combined objectives, such as in agro- or pastoral-silvicultural systems, can be implemented in one program—the very conditions for the forest tree species most often threatened.

It seems clear that combining management techniques will always be necessary for any global program, even for a single species. There are also techniques not easily classifiable as in situ or ex situ but that nevertheless form part of the managerial toolbox for gene conservation. When populations become domesticated and form part of a set of breeding populations, they may change from ex situ status to in situ status if they are allowed to develop and evolve within a new ecosystem. The management of seminatural regeneration, with partial control of parentage or matings between planted and natural stands, is a technique that can be useful for genetic management. It will become more frequently used, at least in temperate forests, in the near future. If in situ conservation stands are too small to ensure their continued evolution, some ex situ stands may be used or developed from the original population and used to regenerate some portion of the in situ stand.

For production forestry, more direct gene management is economically feasible and for some of those species that are already intensively and widely used, advanced breeding programs are in place. For these species, ex situ breeding populations may exist in seed orchards. Supplemental populations may have to be made available, however, to ensure the viability of breeding populations for future uses, and those populations can come from either ex situ or in situ conservation stands or collections. The supplemental populations may be bred for enhanced performance for more widely varying environmental conditions (e.g., the predicted global climatic change), or for more extreme trait expressions than are currently needed. If even these population sets do not satisfy all of the potential needs for breeding or other uses, or if wider samples are desired for saving low-frequency alleles or alleles at risk for other reasons, then additional populations may be maintained in conservation stands or in stands selected for a wider array of uses. Conservation in situ, however, may require more stands than would be required for efficient sampling of the current diversity. If global climatic change is as rapid as some predict it will be, ex situ seeds or stands might be the only source of materials available for breeding.

For several hundred other species that have been described, or are being identified, as potentially useful for production forestry, in situ conservation methods are largely being used—if any conservation programs exist at all. For such species, ex situ stands are being developed in research and testing programs. Hence, heavy reliance must still be placed on in situ conservation until a sufficient array of reproductively competent ex situ conservation stands can be secured. Although this management approach would primarily involve tropical and subtropical species, many temperate species are appropriate candidates for similar management.

The movement of germplasm from in situ to various forms of ex situ use and conservation can be formalized in provenance testing, but this approach has not been effectively developed for large numbers of species of potential but still unknown value. Some organized systems for broad-scale provenance testing have been instituted by such organizations as the Oxford Forestry Institute and the FAO, but a global strategy for efficient, low-cost, and rapid screening of hundreds of species has not been developed.

A nursery worker in Indonesia holds a kadam seedling on her fingertip. About 700 ha of this local species will be planted and later harvested for plywood and furniture. Credit: James P. Blair ©National Geographic Society.

MANAGEMENT STATUS OF TREE SPECIES

The committee has determined that 400 tree species have been included in either breeding or testing programs (see Appendix A). Nearly 500 potentially useful tree species can be classified using the IUCN guidelines as endangered either in whole or at least in significant portions of their range (see Appendix B). Neither list is comprehensive, but both are sufficiently inclusive to allow a general assessment. The greatest conservation efforts are focused on fewer than 140 tree species of industrial forestry value. Of those, slightly more than half are included only in seed collection stands, which in agronomic breeding terms are equivalent to large, mass-selection populations.

Only about 60 species are included in sufficiently intensive breeding programs that ex situ conservation measures such as seed orchards or conservation stands may be said to exist. An equivalent number are included in test or observation stands of one type or another, but only a few are included in international cooperation programs under some designation other than as a stand that may be harvested. They are the *Eucalyptus* in the projects of the Commonwealth Scientific and Industrial Research Organization, the arid-zone and Sahel projects of the FAO, the tropical pine projects of the Oxford Forestry Institute and the Danish Forest Seed Center, the tropical timber species in projects of the Centre Technique Forestier Tropical, and the legumes of the Nitrogen Fixing Tree Association. These organizations and their programs are discussed in Chapter 5. In all of the ex situ programs, the only storage programs other than those included in test or commercial stands are seed and clone banks that mainly serve short-term seed distribution or medium-term storage needs.

In situ gene conservation projects of the FAO include 57 species (in at least one stand), some of which are included in conservation efforts concerning 81 species listed as endangered by the FAO (Food and Agriculture Organization, 1986). For 27 of those 81 species, however, there are no conservation programs for any part of the species, and all have at least some local varieties that are threatened. The biosphere reserve project of the Man and the Biosphere Program at the United Nations Educational, Scientific, and Cultural Organization may ultimately provide some support for in situ gene conservation, but at this time species lists do not exist for two-thirds of the 266 reserves in the program, and most of the lists that are available are for species located in regions where they may not usefully contribute to gene conservation. Many endangered but potentially useful species must be considered to remain largely outside the reserve areas. Some of the endangered species are included in national parks, but most are not in protected reserves, and hence, in situ conservation programs that are genetically effective are primarily of the managed resource area type.

Of the available techniques for tree genetic conservation, heavy reliance is being placed, by default, on managing ex situ and in situ tree stands. Long-term programs for storage of any other materials are very limited. As long as active interest continues, such stand management might continue, but there are obvious systemwide vulnerabilities to even temporary lapses in funding or control.

For genetic conservation in support of intensive breeding efforts with industrial or agroforestry crops, a broad base of different populations is required to accommodate a variety of performance characteristics and a temporally and spatially varying environment. Many of the 60 or so species under some degree of intensive breeding are currently limited to a very narrow genetic base, and their future breeding can be expected to require expanded base populations. Even such widely used species as *Eucalyptus camaldulensis* (river red gum) and *Pinus patula* are already in this state.

It is surprising that there are no long-term germplasm storage programs (i.e., storage of 50 to 100 years) for any of the high-value species. Long-term conservation requires plans for periodic regeneration of stocks and the development of the longest possible storage time to avoid regeneration problems, hence the interest in cryogenic storage in seed and in vitro gene banks. Use of long-term technologies must also be supported by stable organizational structures, guaranteed as appropriate by bilateral international cooperation and perhaps coordinated by an international organization.

Major reliance is placed by foresters on unimproved stands of wild and slightly improved populations for use and conservation. The extinction of important local forms of *Terminalia superba* (nut trees of the western Pacific) and *Acacia nilotica* (Arabic gum tree) and the inclusion of obviously useful species, such as three *Araucaria* (monkey puzzle) species and several *Pinus* species, on the endangered list indicate that current efforts are not yet sufficient even for the species of recognizably high value.

CONCLUSIONS

The choice of which species to conserve and the methods to be employed are dictated largely by either the development of high-value species for breeding or ecosystem conservation. The development of species of only potential value or for diffuse ecosystem values is largely neglected, as are methodological developments for low-cost management of genetic resources in other than intensive breeding programs.

Many species are being harvested commercially, but they are not managed in even-aged plantations and, hence, are not included in breeding programs even though selection could readily improve their economic utility. The number of species that are of sufficient promise for use in the near future to justify some breeding type of genetic management could easily be twice the current number. Such development for use should be backed by genetic conservation in which both in situ and ex situ conservation programs include samples of populations from diverse sources.

For other species of potential value, only a few hundred are included in at least one test plantation and only a relative few are included in extensive international trials. Most of these are either *Eucalyptus* or *Acacia* species, are from limited samples that are not representative of the current habitat variation, and are distributed to only a few planting sites. Thus, it is misleading to suggest from the information in Appendix A that even those species listed are either well sampled or well tested. In fact, fewer than 50 included in the test category in Appendix A are both well sampled and well distributed. About 200 species of clear potential value should be but are not now included in a designed test, and many more should be managed in multiple conservation stands for observation and testing for possibly intensive use. Unless populations of these species are included in an in situ conservation program, they are immediately vulnerable to at least population-level extinction. The 40 or so species currently included in in situ conservation stands represent a small fraction of potentially useful species, only some of which are under any recognized testing program.

For species of even less obvious immediate value to industry or agroforestry, or of little direct use in other forms of forestry, total reliance is placed on natural systems that can withstand human impact. An unknown number are hopefully well conserved with sufficient genetic reserves in ex situ gene banks, designated parks, or other reserves. There is no genetic targeting strategy for forest trees in these programs and, hence, no effort to include useful genetic variation within any forest tree species.

Even in agroforestry the flow of materials from initial observations through testing and breeding is not coordinated and there is no research on breeding methods to make otherwise "primitive" or new varieties useful for production systems. The need to activate a coordinated and forceful new level of effort to conserve and manage forestry genetic resources is critical. It seems obvious that a mixture of the above-described approaches could be efficiently organized by an international agency and that this should be vigorously promoted immediately in order to capture and maximize the future benefit from the remaining diversity of tree genetic resources.

RECOMMENDATIONS

In the areas of conservation and management, additional and increased efforts are needed.

Increase of In Situ and Ex Situ Programs

In situ and ex situ programs to conserve, manage, and use forest tree resources must be significantly expanded to encompass at least a tenfold increase in the species that are included.

Genetic variation is not well conserved for most species, and efforts to conserve and manage tree genetic resources do not encompass global needs. Deficiencies exist both in information and in the extent of activities. For example, ecogeographic surveys must be conducted to assess the need for management interventions. For many species of recognized potential value, new efforts are needed for ex situ conservation. Thousands of species of yet unknown value will require in situ conservation. In tropical and subtropical regions, where species diversity is greatest, many more species should be conserved in situ.

A Global Data Base

A global data base on the status of tree genetic resources should be established and continuously updated. It should include listings of ongoing conservation activities, breeding programs, test stands, and other activities pertinent to conserving trees.

Assessment of the adequacy of existing efforts to conserve species and their genetic diversity requires access to useful information. A global data base would assist in efforts to identify deficiencies in global activities and to allocate often scarce funding resources.

In Situ Management

Long-term, in situ genetic management plans should be developed, especially for tropical and subtropical species.

In situ management plans must consider the number, size, and extent of reserves needed. They must be large enough to ensure survival of tree populations and to conserve their genetic structure, and they must preserve the independent genetic development of separate populations.

Ex Situ Management

Ex situ stands or planted forests should be developed that can serve as living seed banks, as test and evaluation stands, or as both.

Development and application of technologies for the ex situ conservation of pollen, seed, and tissue cultures, as a supplement to in situ maintenance, should be encouraged.

Ex situ methods include managed stands, and the maintenance of seed, pollen, tissue cultures, or other propagation materials. Managed stands permit the ready availability of seed or trees and allow studies of performance under different environments. They could link conservation activities with efforts to develop and improve trees by providing access to significant portions of the gene pool for a particular species.

Programs are needed both to ensure long-term storage (including cold storage) of tree germplasm and to coordinate efforts to maintain managed stands.

Technologies for storing the seeds of many tree species are available and used. However, the long time that may be needed for growing trees to produce seeds and the potential of environmental loss during a lengthy regeneration cycle can make seed production an expensive and uncertain process. Many species do not survive under conditions of long-term seed storage. For them, tissue culture or cryogenic storage could enable long-term storage. Programs for long-term storage of seed, pollen, or tissues are important adjuncts to managed stands, which can be vulnerable to lapses in funding and control, or to environmentally caused loss.

Education and Training

Education and training of professionals and technicians in forest genetic resource conservation should be expanded to provide sufficient technical and support staff to meet urgent needs that will result from increased activity.

Greater efforts to conserve and develop trees will require a concomitant increase in trained professionals and technical staff. Many of the programs described in Chapter 5 include training activities. However, they cannot meet the needs generated by expanded national and international efforts.

5

Institutions Involved in Managing Tree Genetic Resources

The conservation of forest genetic resources has lagged that of agricultural species. Major international efforts to conserve forest genetic resources began only in the late 1960s, with the guidance and support of the Food and Agriculture Organization (FAO) of the United Nations (UN) and several national institutions, mostly in developed countries. In the past two decades, some progress in this direction has been stimulated by two independent sources of interest: (1) tree improvement programs are increasingly demonstrating that economic benefits can be derived from using genetic resources in production forestry, and (2) growing environmental awareness has led to general recognition of the need to preserve natural diversity because of the deforestation of the tropics on a massive scale and the apparent loss of species.

The preponderance of activities worldwide to conserve forest genetic resources focuses on commercially valuable tree species. The main objectives of many of those programs, however, are the acquisition and distribution of high-quality seeds and tree improvement, not genetic conservation.

Breeding programs, however, depend on an adequate supply of new genetic resources. If future demands are to be met by breeding programs, there is likely to be a need for the availability of a wider gene pool than is currently in breeders' collections. This may necessitate both introducing new material and ensuring the conservation of genotypes, populations, and species that are currently available in collections but are in danger of loss resulting from neglect or from the deletion of those materials in disuse.

A nursery worker in Pakistan prepares seedlings for planting as part of a reforestation project. A wide array of tree genetic resources to meet various growing conditions is important for reforestation and afforestation efforts. Credit: U.S. Agency for International Development.

International efforts to conserve forest genetic resources are primarily those of governmental bodies or institutions. The most active organizations are the Oxford Forestry Institute (OFI) in the United Kingdom, formerly the Commonwealth Forestry Institute; the Centre Technique Forestier Tropical (CTFT) in France; the Division of Forestry and Forest Product Research of the Commonwealth Scientific and Industrial Research Organization (CSIRO) in Australia; and the Danish Forest Seed Center (DFSC), which is part of the Danish International Development Agency (DANIDA). The international activities of these organizations, and many others, are supported and coordinated by the FAO's Forest Resources Division. Guidance in planning international activities is provided by the FAO's Panel of Experts on Forest Gene Resources.

Several governmental agencies, especially in the developed world, are also active in conserving forest genetic resources within their national boundaries. The Swedish National Forestry Gene Bank, for example, was established in 1980 to conserve, in perpetuity, a number of indigenous provenances. Younger stands are also to be registered as sources of breeding material for the national program. In general, national efforts to conserve forest genetic resources are usually one component of a larger tree improvement program. The latter programs are designed to meet national goals, as would be expected, and concentrate on commercially valuable species.

In parts of the world where private ownership of forests is widespread, forest industries are involved in preserving gene resources. This is especially true in Central America and parts of South America, where a cooperative effort is headed by the Central America and Mexico Coniferous Resources Cooperative (CAMCORE). This organization, composed of both governmental bodies and private industry, was established in 1980 to prevent the reduction and eventual loss of proven and potentially valuable coniferous species in Central America and Mexico. Field crews in Honduras, Guatemala, and Mexico have collected seeds of endangered species and provenances and have distributed the seed to establish ex situ conservation plots and provenance trials on lands of the members.

In general, the movement of forest genetic material has not been "south to north" as with agronomic crops and about which conflict has arisen over inequities in resources and their technical development (Kloppenburg, 1988). Most exchanges of forest material have been within the temperate zone or within semitropical regions. However, conflicts over the availability and use of forest genetic resources across regional boundaries are likely to arise with the commercialization and internationalization of tropical forestry, the global effects of tropical forest destruction, and the possibilities of global climatic change.

With commercialization will come the potential for disagreements over the equitable distribution of profits between the countries of origin and multinational industries. Further as the loss of forest is increasingly viewed as affecting all nations, the issues of conserving forests in individual countries could become a source of disagreement, particularly where those land areas are viewed as potential sites for agriculture or mining, as has been the case in Brazil. Finally, many of the scenarios for global climatic change portend severe adverse consequences for the world's forests. Addressing this, particularly in developing countries, could become another area of debate.

IN SITU CONSERVATION ACTIVITIES

The extent of in situ conservation of forests worldwide is very difficult to estimate. Very few natural areas have been set aside specifically for genetic conservation, but areas set aside for other conservation needs can serve that purpose. In the United States, for instance, the National Park system, the National Forest Service's wilderness areas, various state parks, and private reserves protect forest ecosystems, which in turn, protect portions of the genetic resources of the contained species. Similar areas have been preserved in many other countries. Most are in the temperate zone, but a growing number of natural areas are being set aside in the tropics. Conserving, managing, or increasing the genetic diversity within species is rarely a direct objective of the programs, however, and thus their adequacy for such purposes is seldom evaluated.

Many of the activities affecting in situ conservation of forest genetic resources have been initiated by the Man and the Biosphere Program (MAB) of the United Nations Educational, Scientific, and Cultural Organization (UNESCO), the International Union for the Conservation of Nature and Natural Resources (IUCN), and the FAO. The FAO's Panel of Experts on Forest Gene Resources has been instrumental in identifying the need for in situ conservation of species used for the production of wood and wood products (Food and Agriculture Organization, 1969). The Panel of Experts has recommended the development of guidelines for in situ conservation (Food and Agriculture Organization, 1972, 1974) and has drawn up operational priorities (Food and Agriculture Organization, 1977, 1985c).

To encourage and stimulate more field projects on in situ conservation, the United Nations Environment Program (UNEP) and the FAO organized an expert consultation on in situ conservation of forest genetic resources in 1980 (Food and Agriculture Organization and United Nations Environment Program, 1981). The purpose of the group was to provide advice on guidelines for the selection and management of in situ genetic conservation areas, the possibilities of combining general conservation with other management objectives (such as ecosystem conservation and production forestry), and needed international actions. One of the recommendations of the group was a project for the preparation of a practical manual on in situ conservation of within-species genetic diversity. This led to the publication of *A Guide to In Situ Conservation of Genetic Resources for Tropical Woody Species* (Food and Agriculture Organization, 1984b).

Several international agencies, such as the FAO, IUCN, UNEP, and UNESCO, have also carried out studies aimed at outlining a methodology for in situ conservation and at drawing up tentative guidelines for action. In addition to the manual mentioned above, two more manuals were published: *In Situ Conservation of Genetic Resources of Plants: The Scientific and Technical Base* (Food and Agriculture Organization, 1984a) and *In Situ Conservation of Wild Plant Genetic Resources: A Status Review and Action Plan* (Food and Agriculture Organization, 1984c). The three manuals summarize what is known of in situ forestry conservation and describe many of the elements that are necessary for successful conservation.

Project plans exist for gene conservation in three countries—Cameroon (Food and Agriculture Organization, 1985d), Malaysia (Food and Agriculture Organization, 1985e), and Peru (Food and Agriculture Organization, 1987). The plans summarize the current status, problems, and opportunities for in situ conservation of 20 important forest species in each country. Initial funding has been obtained, but extensions to other test projects are not planned at this time, and no mechanism exists for carrying the lessons learned from these pilot projects into larger or more comprehensive regional programs. Additional projects exist, such as one operated in Cameroon by the World Wide Fund for Nature (known in the United States as the World Wildlife Fund). Much further testing and development of pilot projects are needed, however, and means must be devised to develop regional programs that complement the efforts of local people and organizations.

EX SITU CONSERVATION ACTIVITIES

Efforts at forest ex situ conservation range from small stands for seed collection, to stands for establishing breeding populations, to international provenance testing programs. Currently, seed stands have been established for some 130 species; breeding stands have been established for 60 species; and international testing programs have been established for some 40 species (excluding species of *Eucalyptus* and *Acacia*). Ex situ efforts are concentrated on economically valuable, fast-growing plantation species. There is little deliberate ex situ conservation of noncommercial species. The CTFT estimates that in West Africa alone 200 humid-zone species exist that have the potential to be useful timber species for plantation forestry, but no programs of collection or testing exist for them. Other nontimber but economically useful species exist; more than 300 species for which several thousand provenances have been sampled by the CTFT are considered useful for some purposes and may be suitable for plantation forestry, but no general conservation program exists for them. Because further research and development work would be required to verify the economic use of these species, the private sector cannot be expected to develop conservation programs, and because national governments cannot afford the expenditure, international efforts are needed to conserve them.

In the United States, about 6,000 ha of established seed orchards and clone banks provide extensive ex situ conservation of some of the most valuable domestic species (U.S. Forest Service, 1982). The thousands of hectares of provenance and progeny test fields and many commercial plantations could also be important components of any coordinated conservation program.

The ex situ conservation programs in tropical countries are, for the most part, smaller and more recent than those in temperate countries. An FAO and UNEP project for the conservation of genetic resources of selected forest tree species and provenances, begun in 1975, established about 40 ex situ conservation and selection stands (about 10 ha each in six tropical Asian and African countries) using 11 provenances of four species—*Eucalyptus camaldulensis* (river red gum), *E. tereticornis* (forest red gum), *Pinus oocarpa* (West Indian pine), and *P. caribaea* (Cuban pine)

These tiny, 8-week-old slash pine seedlings are part of a study to improve the tree stocks of the National Forest system in the United States. Credit: James P. Blair ©National Geographic Society.

(Food and Agriculture Organization, 1981). Many tropical countries are now establishing provenance trials for fast-growing plantation species, and a large proportion of those plantings will probably evolve into seedling seed orchards or provenance conservation stands. In this way, the extent of ex situ conservation will be greatly expanded.

Efforts at forest ex situ conservation also include seed storage at international and national centers. Appendix C contains a partial list of such centers. Very few of the facilities are for long-term preservation of germplasm; many are primarily designed for short-term storage to establish provenance trials and ex situ conservation stands and to distribute high-quality seed. Most national centers are interested primarily in genotypes of local importance; conservation of all possible genes is more important to international institutions. One conclusion that is inescapable is that long-term seed storage of tree species currently plays a very minor role in conservation efforts. This must be rectified quickly.

The FAO has begun a project to assist countries of the Sahelian and north Sudanian regions in initiating research and development activities in the field of genetic resources of woody species. Most specifically, the aim is to enhance or create national seed centers to meet internal demands for reproductive materials by development projects. The project will also set up research programs for the identification, conservation, and improvement of the most promising species. Each participating country will be able to draw on all available genetic variability found within the species it wishes to improve. This implies joint exploration and the organization of seed exchanges between countries in which the species are found.

INTERNATIONAL INSTITUTIONS

This section presents information on selected international institutions working in the areas of forest conservation and research, with varying levels of involvement, specifically with regard to forest genetic resources. The information illustrates the scope and diversity of activities undertaken by international institutions to manage forest genetic resources. Because most of the institutions mentioned contribute to the area of forest genetic resources management but may not be specifically involved in the field, the budgetary information presented, which was drawn from annual reports, trip reports, and interviews, is often sketchy. The descriptions of institutions and projects are presented as outlines of the known activities of each institution. A detailed analysis of the functions and impact of the institutions would require a separate, intensive examination, a task beyond the scope of this report. The coverage is intended to provide a global overview of current activities and of worldwide needs that have yet to be addressed.

Food and Agriculture Organization

The FAO, headquartered in Rome, plays an important role in coordinating and implementing forest genetic resource policy within its overall aim of providing technical assistance. The FAO's coordination activities are conducted through the work of its Panel of Experts on Forest Gene Resources, and its planning activities are conducted through its Ecosystems Conservation Group and its Working Group on In Situ Conservation of Plant Genetic Resources (members are the FAO, IUCN, UNEP, and UNESCO). The FAO's forest genetic resources officer also

One of the world's most dramatic erosion and deforestation situations is in Nepal, where forests up to elevations of 2,000 m have totally disappeared. Concerted efforts have taken place in many areas to provide farmers with their own supplies of animal fodder and fuelwood to reduce pressures on natural forests. Here members of a village community in Nepal meet with a forest extension agent to discuss the planting of quickly maturing trees. Credit: Food and Agriculture Organization.

participates as a member of UNESCO's Scientific Advisory Panel for Biosphere Reserves (Palmberg and Esquinas-Alcázar, 1990).

The FAO's forest program has placed increased emphasis in technical terms on the use of multipurpose species for the provision of a range of goods and services and for environmental rehabilitation. Increasing amounts of funding for this purpose are being channeled through national institutes in developing countries. In the area of genetic resources conservation, increased attention has been paid to in situ conservation as a desirable complement to various forms of ex situ conservation.

Because of the high human and biotic stress to which they are subjected, species in the arid and semiarid regions of developing countries are given priority in the FAO's programs. To satisfy both immediate and future needs, the FAO emphasizes locally controlled activities and training. Thus, it favors projects that rely on national or local expertise, for example, to collect and establish test stands for breeding and conservation (in contrast to the OFI, which has its own professional staff and focuses on international sampling collections).

The FAO's Panel of Experts on Forest Gene Resources was established in 1968 to advise the FAO on its efforts to explore, use, and conserve gene resources of forest trees and, in particular, to help prepare short- and long-term programs for FAO's action in this field and to provide information to member countries. The panel has met six times (1968, 1971, 1974, 1977, 1981, 1985) in 17 years to review the current extent of the depletion of forest genetic resources, to make recommendations for methods of conservation, and to establish priorities for action and international support.

At the recommendation of the Panel of Experts, the FAO began to publish a newsletter, *Forest Genetic Resources Information*, in 1973. The newsletter disseminates information on forest tree seed supplies; seed collection, handling, storage, testing, and certification; the organization and results of international provenance trials; and various aspects of the conservation and use of forest genetic resources. The newsletter is published in English, French, and Spanish and distributed free of charge to state forest services, forest research institutes, universities, and individual scientists who have expressed an interest in the publication. The FAO (1986) has also produced a *Databook on Endangered Tree and Shrub Species and Provenances*. Eighty-one species, endangered either at the species or provenance level, are the focus of the book. Its main purpose is to draw the attention of decision makers, scientists, and international and national organizations to the need to conserve the species in question.

Another major activity of the FAO is the support of seed collection and the handling, seed storage, and evaluation of seedlots. The DFSC, among other seed centers, assists the FAO by providing short-term seed storage and distribution facilities for international use. It was originally created at the recommendation of the FAO Panel of Experts.

With the Forest Resources Development Branch, the FAO's core program in forest genetics has limited but constant funding, which annually is about 6 percent, or $100,000, of the branch's overall budget. This does not include salaries or funds to support extrabudgetary and field projects. Genetic resources activities are also conducted by other FAO entities, such as the Forest Wildlands and Conservation Branch and the Plant Production and Protection Division. The division of labor among the various FAO entities with responsibility for forests and trees seems to be well defined and to minimize overlap.

The FAO's recently established Commission on Plant Genetic Resources is intended to provide policy guidelines for FAO activities in the general area of plant genetic resources, including forestry. The Forestry Resources Division would then implement recommendations of the commission regarding forestry.

Another FAO activity that includes a forest genetics component is the Tropical Forestry Action Plan (TFAP) (Food and Agriculture Organization, 1985c). The program, which is intended to provide a framework for global action in tropical forestry, specifies five broad priority areas: forestry in land use, forest-based industrial development, fuelwood and energy, conservation of tropical forest ecosystems, and institutions. The plan for conserving tropical forest ecosystems outlines the desirable actions and funding required for ecosystem conservation and the conservation in situ of inter- and intraspecific variation of genetic resources of target species. In particular, the TFAP calls for (1) carrying out botanical surveys of plant diversity and distribution, (2) developing methods to protect plant diversity and species variation, (3) developing conservation data and increasing awareness of the values of genetic conservation, and (4) increasing research on species of potential economic value. It is still too early to evaluate the impact of this program on global conservation of forest genetic resources. The TFAP has, however, provided a framework within which international donor agencies can focus activities, and at a November 1988 meeting, organizational plans were advanced for research and development programs.

The International Board for Plant Genetic Resources

The International Board for Plant Genetic Resources (IBPGR) is an international, scientific center of the Consultative Group on International

Agricultural Research (CGIAR). It is autonomous and governed by an independent board of trustees. It was founded in 1974 and its headquarters location was provided at the FAO in Rome until its move to separate offices in that city in 1989. For a number of years the FAO provided some staff and subsidized office space to the IBPGR, and during that period the FAO program on crop genetic resources was coterminous with that of the IBPGR. When it was created, the IBPGR had about six national programs with which to work; currently it works with 110 countries. It has been instrumental in the initiation of many national programs and the establishment of gene banks, and it supports training and research.

The original mandate of the IBPGR was broad enough to include woody species. However, the preference of its donors led the IBPGR to focus its efforts on agricultural crop species. It has, however, supported and cooperated with various efforts involving forest species, particularly fuelwood species. Additionally, the priority crops of the IBPGR include a number of woody species (citrus, cacao, breadfruit, oil palm, and *Prunus*) for which a substantial amount of work has been supported. Several of the IBPGR's publications have particular relevance to conserving tree genetic resources.

The IBPGR's field program has an infrastructure of professional staff located at its regional offices for China, south and southeast Asia, west Africa and the Sahel, east and south Africa, Central America, South America, and southwest Asia, the Mediterranean, and Europe. Through them, the IBPGR targets support and advice and organizes priority work on collecting, conserving, characterizing, and storing germplasm and on analyzing data.

Since its inception the IBPGR has evolved a policy and strategy planning capability and an infrastructure related to national programs that could, with adequate funds, provide a rapid start-up for new programs. The IBPGR is considering how it could expand its work into woody species, and develop scientific standards and implement priority programs (van Sloten, 1990).

International Council for Research in Agroforestry

The International Council for Research in Agroforestry (ICRAF) was established in 1977 in response to a series of recommendations in a study undertaken for the International Development Research Center (IDRC) of Canada. Headquartered in Nairobi, Kenya, the ICRAF was established as an international scientific center devoted to improving the nutritional, economic, and social well-being of people in developing countries by promoting agroforestry systems for enhanced use of the land without degrading the environment. The ICRAF is an autonomous agency, governed by an independent board of trustees with equal representation from developed and developing countries. It is expected to be included in the CGIAR system as part of an overall forestry effort (Consultative Group on International Agricultural Research, 1990).

The ICRAF acts as a catalyst for agroforestry research, training, and information dissemination. It has verified, for example, that a very high number of agroforestry systems exist worldwide and that, most critical to tree genetic resources, more than 2,000 multipurpose trees are being used in complex agricultural systems. It has also published a directory of worldwide sources of multipurpose tree seed. This type of information is very important to those using and managing minor forest tree genetic resources (von Carlowitz, 1986).

In 1989, the ICRAF received a total of more than $7.2 million in financial support from governments, foundations, national organizations, and international institutions. Of this, $2.96 million was designated for core program support, and $4.3 million was for project activities (International Council for Research in Agroforestry, 1990).

International Tropical Timber Organization

The International Tropical Timber Organization (ITTO) was mandated by the International Tropical Timber Agreement, 1983, which was formulated at the United Nations Conference on Tropical Timber (United Nations, 1984). It became fully operational in April 1985. The purpose of the ITTO is to provide an effective framework for cooperation and consultation between tropical timber-producing and -consuming countries regarding all aspects of the tropical timber economy.

The principal governing body of the ITTO is a council composed of 42 member governments, which represent 95 percent of world trade in tropical timber and 70 percent of the global tropical forests. There are three subsidiary bodies in the form of permanent committees on Reforestation and Forest Management, Forest Industry, and Market Intelligence and Economic Information. The funding for ITTO's administrative activities is generated from annual contributions by member states (1989 budget of $2.1 million). The organization's projects are funded separately by the voluntary donations of members (1989 project budget of $8.5 million).

The ITTO, based in Yokohama, Japan, meets twice yearly in various locations around the world, including Rio de Janeiro, Brazil; Abidjan, Ivory Coast; and Bali, Indonesia. As a commodity-oriented organization, its activities span the range of trade and industry issues, including maintenance of the tropical forest resource base. The major activity of the ITTO related to forest genetic resources involves funding of projects directed at the conservation and sustainable use of tropical forests. Projects funded have included study of the rehabilitation of commercially logged forests in Asia, research on natural forest management in Malaysia, and research and pilot studies on areas damaged by fire in East Kalimantan, Indonesia. During the Rio de Janeiro meeting, in June 1988, the ITTO approved funding for a study of integrated forest-based development in the western Amazon. This project, the focus of which is sustained forest management in the Brazilian state of Acre, will be conducted in a 100,000-ha forest reserve used by local people for tapping rubber trees and collecting Brazil nuts and other forest products. Similar, smaller scale pilot projects for Bolivia and Ecuador were also approved at that meeting.

Thus far, little mention is made in ITTO's studies of a genetic aspect to managing or conserving forest resources for future industry uses. In most of its efforts, forest management refers to managing commercial tropical timber species with maximized returns and improving harvesting methods. The emphasis on sustainability and forest management is a positive step for this international commodity organization.

International Union for the Conservation of Nature and Natural Resources

The IUCN, headquartered in Gland, Switzerland, is a unique international agency in that it was constituted with both a governmental and nongovernmental membership. It was created in 1948 with the support of UNESCO. In the past decade, the IUCN has become concerned with species-level conservation and has paid increasing attention to plant genetic resources. Funded by the UNEP and the World Wide Fund for Nature, the IUCN prepared the World Conservation Strategy, with the technical assistance of the FAO and UNESCO (International Union for the Conservation of Nature and Natural Resources et al., 1980). Measures to protect genetic and biological diversity were a major component of the strategy.

The IUCN's budget for 1988 was about $14 million, half of which was allocated to biodiversity projects (e.g., parks management and conservation monitoring centers). The organization has a variety of expert commissions and committees that address aspects of biological conservation. Its Threatened Plants Committee, set up in 1974, produces the IUCN *Plant Red Data Book*, a series that gives detailed case histories (red data sheets) on rare and threatened plants in all parts of the world. For each species, data are given on conservation status, threats to survival, distribution, and habitat, together with a short description and an evaluation of its interest or potential value to humankind.

The IUCN has a World Conservation Monitoring Center (WCMC), located in Cambridge, United Kingdom, for data storage and processing. The WCMC also provides information on international trade in endangered plants and animals, especially through the Convention on International Trade in Endangered Species of Fauna and Flora. Another unit, the Protected Areas Data Unit, deals with data on ecosystem conservation. At present, the units deal with biological resources, mostly at the species level and sometimes at the habitat or ecosystems level. They do not maintain data on genetic variation within species.

International Union of Forestry Research Organizations

The International Union of Forestry Research Organizations (IUFRO), headquartered in Vienna, Austria, does not directly conduct programs but coordinates and assists scientists participating in programs. It has traditionally placed strong emphasis on industrial species and has several working parties on provenance testing, progeny testing, and breeding of specific species. It also has a working party on conservation and one on population genetics. Included in these efforts are several temperate-zone and Mediterranean conifers, as well as *Quercus* (oak), *Eucalyptus*, and *Populus* (poplar) species. Recently, a very active working party on tropical species was formed. Material and information on those species of clear and high potential value for production forestry are provided to interested parties at a low cost. The IUFRO's total budget for 1987 was slightly more than $234,600, about 65 percent of which came from membership dues.

The IUFRO's Special Program for Developing Countries, established after the organization's 1981 world congress, has organized four regional research planning workshops. During these workshops, the conservation and use of genetic resources was recognized by donors and developing country representatives as an issue for priority attention.

Following the first IUFRO planning workshop, for Asia in 1984, the U.S. Agency for International Development (USAID) started a project on fuelwood (Fuelwood/Forestry Research and Development) that includes the collection and dissemination of information on species for wood energy (International Union of Forestry Research Organizations, 1985). The second IUFRO planning workshop, for the Sahelian and north Sudanian regions of sub-Saharan Africa in 1986, eventually led to the formulation of national genetic resource projects concerning

Logs of Philippine mahogany, a species of dipterocarp, are being transported from a rain forest on the island of Borneo, Indonesia. They bring high prices in the international lumber market. If they are not carefully harvested, significant damage to the forest can occur. Credit: James P. Blair ©National Geographic Society.

multipurpose woody plants in the Sudano-Sahelian zone. The projects were coordinated through a FAO project financed by a French trust fund (International Union of Forestry Research Organizations, 1986, 1987). The third workshop in 1987 focused on the role of multipurpose species in community forestry in Latin America and the Caribbean. The fourth workshop in 1988 identified needs and prepared collaborative projects for genetic evaluation and breeding methods for southern and eastern Africa using many indigenous species and some exotic species.

United Nations Environment Program

The UNEP, headquartered in Nairobi, Kenya, is an agency of the UN. It was established in 1973 and charged with working with governments, other UN organizations, and nongovernmental organizations around the world to monitor the state of the global environment. The UNEP essentially provides a catalyst for actions to be taken to address the resource conservation needs of member nations. Its actions are largely undertaken in collaboration with other UN agencies, such as the FAO, but it works closely with organizations outside the UN umbrella, such as the IUCN and the IBPGR. It provided, for example, the funds for a FAO project on forest genetic resources that led to the establishment of two in situ genetic reserves in Zambia and several ex situ conservation stands in tropical Asia and Africa; to the production of the report *Methodology of Conservation of Forest Genetic Resources* (Roche, 1975); and to the expert consultation on in situ conservation of forest genetic resources (Food and Agriculture Organization and United Nations Environment Program, 1981).

Much of the UNEP's work is aimed at promoting public awareness of the importance of genetic diversity and methods of conserving and managing that diversity for the future. The UNEP also acts as a facilitator. In 1985, it invited the African Ministers Conference on the Environment to hold its initial meeting in Cairo. At that meeting, it was decided that an African Genetic Resources Network would be created to promote the conservation and management of the biological diversity within African countries. The conference also decided to create a regional committee on forests and woodlands to advise members on the status of forestry and forest products in the region.

The UNEP was initially scheduled to receive about $35.5 million in financial support for 1988, to which $500,000 was later to be added. About 50 percent of the funds were allocated to activities directly implemented by the UNEP, and the remaining funds were about equally divided between support of other UN activities and of governmental and nongovernmental programs.

United Nations Educational, Scientific, and Cultural Organization

UNESCO's involvement in gene conservation is primarily through its MAB program. The objectives of MAB have evolved from establishing project areas based on an ecosystem concept (including human activity), to conserving representative ecosystems with zoned management, to developing biosphere reserves that conserve biological diversity and its genetic resources. In cooperation with various UN and other agencies (e.g., IUCN, the U.S. Smithsonian Institution), MAB is engaged in a program of inventorying and monitoring all the vegetation in the biosphere reserves and creating a system of inventory that will enable a global estimate to be made of the extent to which biosphere reserves are assisting in the conservation of genetic resources and biodiversity.

The MAB program works through a network of pilot projects—at present mainly in the humid tropics. Its draft program and budget for 1990–1991 amount to roughly $2.7 million, of which about $725,000 is earmarked for in situ nature conservation and another $200,000 for concerted action at national, regional, and international levels. UNESCO, headquartered in Paris, is working with the FAO, IUCN, and UNEP and has combined with them to form an Ecosystems Conservation Group. The awareness of genetic, as distinct from ecosystem, conservation objectives has led to UNESCO's participation in several in situ projects for gene resource conservation with the government of Mexico and the IBPGR, among other agencies.

Consultative Group on International Agricultural Research

The CGIAR was established in 1971 to help coordinate the efforts of developed and developing countries, public and private institutions, and international and regional organizations to support a network of 13 international agricultural research centers. It provides a mechanism for obtaining financial support for the centers. Through international agricultural research and related activities, its goal is to develop technology and cooperate with national research systems in developing countries with the aim of alleviating hunger and poverty, improving the management of natural resources, and increasing employment and income. The CGIAR is cosponsored by FAO, the United Nations Development Program, and the World Bank (Baum, 1986; International Board for Plant Genetic Resources, 1989). Its secretariat is located in Washington, D.C.

Plans exist within the CGIAR for expansion to address forestry issues (Consultative Group on International Agricultural Research, 1989a,b; 1990). Discussions have focused on two broad research areas: deforestation, including addressing its causes, amelioration, policy reforms, and conservation of forest resources; and policies and technologies to encourage the activities of reforestation and natural forest management. However, the management of tree genetic resources may form only one aspect of the CGIAR's effort, which will be implemented through ICRAF and a newly created center. Nonetheless, the effect of increased activities in areas such as silvicultural systems, agroforestry, and improvement of multipurpose tree crops would be to heighten both the interest in and urgency for the conservation, management, and use of tree genetic resources.

The CGIAR could become involved in a global effort to manage tree genetic resources: (1) through the IBPGR, as described above, (2) through a cooperative activity among several of its institutes, or (3) through a new center established for this purpose. Of the options above, involvement through the IBPGR would seem to be the most appropriate for those aspects related to conserving and managing tree genetic diversity.

NATIONAL AND REGIONAL INSTITUTIONS

This section describes selected national and regional institutions with international programs bearing on forest genetic resources. The information presented, as with the international institutions, is a sampling of the approaches and programs now in operation. The information may be of use to countries wishing to establish or expand their work in forestry genetic resources. Complete coverage of the activities and budgets of these organizations has not been attempted.

Central America and Mexico Coniferous Resources Cooperative

Cooperative efforts such as CAMCORE are one of the few ways in which private organizations have been able to affect directly the development of genetic resources. The CAMCORE is composed of 13 active cooperating entities in nine countries; about half are private companies and the remainder are units of governmental research and tree breeding agencies. Each entity contributes to the support of the activities of the cooperative and also contributes in-kind services to local collection and planting projects. In the United States, USAID and the Rockefeller Foundation have provided grant funds for collection projects. Joint efforts are also undertaken with other agencies, such as the Banco de Semillas Forestales in Guatemala, the Escuela Nacional de Ciencia Forestal in Honduras, and Centro Agronómico Tropical de Investigación y Enseñanza (CATIE) in Costa Rica. Program funds for 1988 were about $250,000, all of which was directed to forest genetic resource programs.

The CAMCORE has sponsored collections for some 30 Central American and Mexican conifer and angiosperm species. From the collections, seeds are processed, stored, and distributed to cooperating agencies for establishing conservation stands and provenance and progeny tests. The cooperative helps set goals and design studies, employs a headquarters staff and some field staff, organizes and oversees field activities with the assistance of local cooperators, and analyzes the test data. Each cooperating entity provides its own planting and testing facilities to the extent it chooses, in accord with its potential use of the materials for supporting local breeding efforts and for exchanging materials and data.

As a volunteer cooperative, the CAMCORE depends entirely on annual contributions; it also enjoys the support and carries the academic credentials of North Carolina State University, which lends stability as well as access to scientific support. A balance is, therefore, struck between satisfying the somewhat longer-term interests of an academic community and those of the governmental entities. Thus far, support for the cooperative has been stable and growing.

Centro Agronómico Tropical de Investigación y Enseñanza

In 1967, the Forestry Department of CATIE founded the Latin American Forest Seed Bank. Its establishment was due mainly to the interest of several Latin American countries in introducing exotic species to meet the forest industry's demand for wood. A number of such projects were once operational in Central America. Annually, the CATIE collects and distributes seeds of about 50 species to 35 countries for plantation and experimental purposes. The member countries of the CATIE are Costa Rica, Dominican Republic, Guatemala, Honduras, Nicaragua, and Panama.

One of the CATIE's technical activities is the Renewable Natural Resources Department, which has four programs: agroforestry, silviculture, wildlands, and watershed management. The agroforestry and silviculture programs involve tree improvement and ex situ conservation. One of the primary objectives of the wildlands program is managing protected areas in Central America and elsewhere in Latin America, including in situ genetic resource conservation. The wildlands program assists in identifying and developing national parks and protected areas within Central America. It also acts as a regional liaison for international organizations interested in conservation in the region. It has carried out exploration and collection of *Gliricidia sepium* and *Calliandra calothyrsus*, and is promoting the establishment of seed stands of certain species within countries of the region in conjunction with a tree crop production project.

The CATIE has suffered from insufficient and insecure financing ($10 million to $12 million annually). It possesses, however, the necessary basic infrastructure to give important support to the conservation of forest genetic resources and, consequently, forestry development in the countries of the region. It is in a strategic location to create and manage in situ conservation reserves and to train future forestry scientists.

Canadian International Development Agency and International Development Research Center

Canada provides substantial assistance for forestry research in developing countries through projects financed by donors such as the Canadian International Development Agency and the IDRC forestry

A peasant carries wood from a section of rain forest in Zaire that is being cleared and burned. Agroforestry and reforestation programs can reduce forest losses and restore degraded sites. Credit: James P. Blair ©National Geographic Society.

program. The IDRC's program covers reforestation, production systems and agroforestry, use of forest products, tree improvement, and conservation. The IDRC also organizes seminars and training courses and supports the IUFRO's research planning workshops. Half of its forestry effort is devoted to Africa, a quarter to Latin America, and a quarter to Asia.

A major Canadian initiative in the area of genetic diversity is the joint establishment of a facility in Thailand by the Association of Southeast Asian Nations (ASEAN) and the Canada Tree Seed Center to assist the ASEAN members in developing seed technology relevant to forest renewal in southeast Asia. Concern for genetic resources and their maintenance in situ and ex situ are critical components of the seed center's program. The IDRC in general does not carry out projects; rather, it provides support to scientists in developing countries to carry out projects that match the aims of its forestry development program. The IDRC's projected budget for 1988–1989 was about $120 million, 95 percent of which was a grant from the Canadian government.

Centre Technique Forestier Tropical

The CTFT is a department of the French foreign assistance agency Centre de Coopération Internationale en Recherche Agronomique pour le Développement (CIRAD), which has its origins in various tropical forestry agencies of the French government dating back to 1917. The CIRAD is a large organization of about 1,200 people; the CTFT has about 200 professionals. The CTFT collaborates with the FAO, from which it receives some support.

The CTFT's style of operation is very intensive and narrowly focused. The CTFT concentrates, for example, on 12 francophone sub-Saharan African countries plus French Guiana and New Caledonia. The CTFT also focuses its activities on only a select number of species: *Eucalyptus* species for the humid and dry tropics, *Pinus caribaea* and *P. kesiya*, *Terminalia superba* and *T. ivorensis* (nut trees of the western Pacific and Africa), *Tectona grandis* (teak), *Gmelina arborea* (used for timber in tropics), *Cedrela odorata* (West Indian or Spanish cedar), *Cordia alliodora*, and *Acacia mangium*, *A. auriculiformis*, *A. albida*, and *A. senegal*. The CTFT's activities with these species are very intensive; they deal with all aspects of forestry operations, from development of breeding and nursery techniques to primary conversion and utilization.

Species priorities vary widely among the tropical countries with which the CTFT has bilateral agreements. In general, species priorities are determined by economic importance, and the interest in genetic resources is based on use. Some short-term conservation objectives are met by the CTFT's stand management project, and some medium-term objectives are met by its seed collection activities, but the CTFT is not engaged in any long-term in situ or ex situ projects.

The CTFT has developed seed storage, research, and facilities for supplying seed of highly valuable species for evaluation plantings and other needs in nations with which it has cooperative projects. It also has test plantings of its target species in several countries. Although the CTFT is aware of the need to ensure the continued existence of at least a sample of these plantings and also to establish in situ conservation stands, it has no long-term genetic conservation projects.

The CTFT's 1988–1989 budget for genetic resources was $825.7 million. About $220.2 million of that amount was earmarked for particular project categories: collection and evaluation of tropical forest genetic resources ($86 million), conservation in the tropical zone ($65.4 million), and laboratory research and studies ($68.8 million). The funds are provided by the French government and European communities.

Commonwealth Scientific and Industrial Research Organization

In Australia, forest genetics work is conducted through CSIRO's Division of Forestry and Forest Products Research, Program on Australian Tree Resources, Canberra. About 75 percent of the CSIRO's funds are appropriated by the Australian Parliament; contributions by industry and other groups account for most of the remainder. Total research expenditures by the CSIRO in 1988 amounted to roughly $362.6 million, of which nearly 23 percent ($82.8 million) was allocated to plant production and processing, the category into which forest genetics activities likely fall. The committee was not able, however, to obtain a more precise estimate of CSIRO's expenditures related to forest genetic resources.

The Tree Seed Center, a key unit of the Program on Australian Tree Resources, is a focal point for many projects that require access to Australian forest genetic resources for use in other countries. The center collects and distributes high-quality, source-identified seed of commercially promising Australian woody plants for research purposes, provides professional advice on the choice of species and seed supply, and provides technical information on species of value and makes the materials widely available.

During 1982–1985, the center undertook 45 major collecting programs, about half of which involved the collection of *Eucalyptus* seed. The major activity focused on individual tree collections from superior provenances of proven species for tree breeding purposes. One project, Australian Trees and Shrubs for Fuelwood and Agroforestry, funded by the Australian Center for International Agricultural Research (ACIAR), has meant a new direction for the center. This project involves the exploration and collection of seed of lesser known Australian trees and shrubs with potential for use in agroforestry and fuelwood plantings in developing countries. The program supports conservation of species both directly through seed storage and ex situ plantings and indirectly by drawing attention to their potential for utilization. It also provides a stimulus to sample genetic resources of non-*Eucalyptus* species. The center's work in this area has concentrated mainly on nitrogen-fixing *Acacia* and *Casuarina* (she oak) species.

The center collaborates with Australian and international donor organizations in arranging and distributing the seed for international provenance trials of important species. Trials of *Acacia mangium* and *A. aneura*, for example, have been established with the participation of 40 organizations and 11 countries. The CSIRO and the Queensland Forestry

Department, with support from the ACIAR, are developing a data-base system for storing and selectively retrieving the results of field trials, especially those using well-documented seedlots provided by the Tree Seed Center. Information on the environment of the test site and on the management practices used is also included.

The CSIRO has noted that there is less pressure on natural stands of economically important species in Australia than in other countries, but in situ reserves are still established as part of large-area reserves managed by the National Park Services. The CSIRO has also noted that a major constraint to the most effective conservation and use of Australian genetic resources in other countries has been the lack of a recognized international framework for such activities.

Danish Forest Seed Center

The activities of the DFSC, a forestry development assistance organization financed by DANIDA, were initiated in 1969 under the project name Danish/FAO Forest Tree Seed Center. The DFSC is based in Humlebaek, Denmark, although most of its activities take place in developing countries, primarily in southeast Asia and Central America. Procurement of large quantities of high-quality seed is one of its main objectives. The DANIDA has been focusing on three important species: *Tectona grandis, Pinus merkusii,* and *Gmelina arborea.*

A major activity of the DFSC has been acting as a handling, storage, and distribution point for seed collected within the FAO/IBPGR/UNEP project, Genetic Resources of Arid and Semi-Arid Zone Arboreal Species for the Improvement of Rural Living. A related activity has been assisting other national programs in the evaluation of international provenance trials coordinated by the DFSC, which includes field measurements, computation of data, and interpretation of results. This is regarded as one of DFSC's highest priority activities.

DANIDA has also been providing assistance to several countries for the management of ex situ conservation stands to help maintain the greatest possible genetic diversity within the stands. Collections of *Pinus kesiya* and international provenance testing have been carried out in The Philippines, Thailand, and Vietnam in cooperation with the Oxford Forestry Institute.

In 1980, DANIDA spent just under $450 million on forestry projects, about 7 percent of its total budget. Funds are provided by the Danish government and then channeled into bilateral and multilateral projects with the UN's member nations and organizations.

Oxford Forestry Institute

The OFI is a world center for forest research and development. The need for adequate supplies of correctly named, site-identified seed trees grown for industrial and nonindustrial purposes led the OFI to establish international provenance testing projects for some 50 species in Central America and parts of Africa. The projects cover exploration, taxonomy, collection, seed storage and distribution, field trials, establishment of conservation stands, evaluation, and conservation and development of genetic improvement strategies for a number of tropical species.

Since 1963, the OFI has made collections in the entire Central American region for provenance and progeny trials, especially of pines. Since 1980, it has also made collections of tropical broadleaf trees from arid and semiarid zones in Central America. Seed from OFI collections are distributed free of charge for trials all over the world. The evaluation of such trials has already provided a great deal of information about genetic variation within the collected species.

The emphasis in all OFI genetic projects is on breeding for use in the tropics. The species OFI initially worked on were largely tropical pines, *Pinus caribaea*, *P. oocarpa*, and *P. tecunumanii*. Recently, it has expanded its work to include *P. patula*, *P. kesiya*, *P. merkusii*, and *P. greggii*, among the pines, and to a lesser extent, other gymnosperms, including *Agathis* (kauri), *Cupressus* (cypress), and *Widdringtonia* (African cypress) species and *Abies guatemalensis*. Among the hardwood species with industrial uses, projects are being developed for *Cedrela* (Chinese cedar) species, *Cordia alliodora* (Ecuador laurel), and *Liquidambar styraciflua* (species of sweet gum) from Central America. For agroforestry use, *Gliricidia sepium* and several species of *Leucaena* and *Prosopis* are included in initial species assays, and in east Africa several species of *Acacia (A. albida, A. tortilis, A. senegalensis*, and *A. nilotica)* are being collected in various cooperative projects. Some 25 tree species from the Central American dry zone, mainly legumes, are also being studied.

For all of the above species, a two- or three-stage sequence of projects is followed. The first stage begins with initial collection of materials for identification and study of phenotypic variation. A second stage is a more formal ecogeographic study using rangewide collections and widely distributed trial plantations to establish initial provenance trials. A third stage is sometimes necessary for more intensive sampling of selected individuals or populations. Selections from the initial provenance trials are also used with materials from the third stage to establish local breeding populations.

The models of the OFI's operational style are its tropical pine projects (*Pinus caribaea, P. oocarpa,* and *P. tecunumanii*) mentioned above. The initial provenance trials included some 30 sources, planted in several hundred plots, of which about 15 sites are considered representative and reliable for detailed data analysis over several environments. About one-third of the original sources were considered useful for resampling, and with substantial intrapopulational genetic variation and at least 25 trees sampled per population, a sufficient breeding base is considered to have been established. Several breeding programs have been developed from this start.

The impact of these projects on the genetic resource base is mixed. *Pinus tecunumanii* has been well sampled and exists in enough diverse test or breeding populations that its genetic variation is currently well conserved. *Pinus patula, P. caribaea,* and *P. greggii* are well sampled but a few populations are missing from collections. There are more gaps in *Pinus oocarpa* and in most of the commercial hardwood collections, especially of populations at the edges of species ranges, where many local variants are being lost or are uncollectible.

The OFI has its own in-house technical staff (and other professionals on outside funding) but the institute largely depends on short-term grants and, hence, funding becomes a constraint on program continuity. There are 11 professionals in the Genetics and Tree Breeding Unit, all on non-university-funded ("soft") projects. The unit is responsible for all projects on genetic resource development. It cooperates with many national governments and international agencies, has set up seed centers in developing countries, has used those seed centers in its genetics projects, and has consistently made all materials and information freely available to all. Considering that it costs roughly $1 million to establish the pine trials by its standard procedure, and that its funding cycle is for 2- or 3-year projects, the OFI takes a very intensive but short-term view of project priorities. Building infrastructure and developing personnel in the countries where it works is not a major component of a project, and hence, the OFI does its own seed collection and distribution. One of its other functions is, however, academic; in particular, the OFI offers intensive courses in Oxford and abroad in a wide range of subjects.

U.S. Department of Agriculture's Forest Service and State Programs

The activities of the U.S. Department of Agriculture's Forest Service include research and genetic improvement programs conducted in national forests. Research is primarily aimed at species of high commercial value, but there is a growing emphasis on maintaining diversity in forests. The Forest Service also provides international assistance in forestry to USAID's overseas missions.

The Forest Service has collected and maintained varieties of *Pinus taeda* (loblolly pine) since 1943. Storage facilities for tree germplasm are located at its laboratories. These are not long-term storage facilities, however, and there is no national inventory of what tree germplasm is in storage. When the Forest Service identifies endangered populations of forest tree species, it collects germplasm and, when possible, grows it at a site with similar environmental conditions. Most U.S. national parks have set aside land as genetic management areas.

Individual states also have independent conservation programs, and several states have a tree improvement program that includes genetic resources components. State activities generally do not have an international aspect, but they can serve as models for larger efforts. California, for example, has an innovative project to conserve conifer genetic resources. The Conifer Germplasm Conservation Project is designed to provide information and resources needed for long-term protection of the diversity of California's forests. As part of this program a major effort will be to complete seed collections in California from the range of *Pseudotsuga* (Douglas fir) and *Pinus ponderosa* (ponderosa pine). Intensive seed collection also will be made for *Picea brewerana* (Brewer's spruce), a rare and little known species found in a few locations in northern California. Related aspects of the project will include research on the geographic patterns of genetic variation in the above-mentioned fir, pine, and spruce species, restoration of several ongoing gene conservation collections that have fallen into disrepair, and a computer-based catalog of lands dedicated to forest gene conservation.

CONCLUSIONS

Funding levels for forest tree genetic resources programs around the world are inadequate to ensure the continuation of even the current level of activity. Base program funding is being reduced in some organizations and was never adequate in others. Support for tree genetic resources is, therefore, often the product of the personal interests of individual scientists. The situation is exacerbated by the frequent lack of defined policies and programs designed specifically to prevent the loss of valuable genetic resources. It has been roughly estimated (S. Krugman, U.S. Forest Service Timber Management, personal communication, 1989) that the total expenditure worldwide specifically for forest germplasm conservation activities is $5 million annually. If the

Haiti's landscape (left) contrasts sharply with the rich forests of its neighbor, the Dominican Republic. Although 95 percent of Haiti has been deforested, trees continue to be cut down to expand farmland and to produce charcoal. Reforestation efforts have been inadequate. In this case, the magnitude of the problem far exceeds the national resources to address it. Credit: James P. Blair ©National Geographic Society.

rapid rate of loss of genetic resources is to be reversed, there must be a significant increase in this level of funding.

Increases in funds are, however, frequently difficult to obtain. Many valid programs and concerns compete for a limited pool of funding. Although coordination at the international level is a recognized need, the funds available to support such an undertaking may be limited. Thus, programs at the national, regional, and international levels must be designed to build on existing institutional capabilities and to provide the greatest level of activity within possible funding constraints. This may be achieved by setting priorities, networking activities, sharing responsibilities (among national, regional, and international institutions), and integrating activities with other natural resources programs (e.g., those of the IUCN, the World Wide Fund for Nature, botanical gardens and arboreta, and development projects).

The institutions discussed in this chapter are carrying out important work at national and international levels. However, gaps remain in both priorities for and methods of conserving forest genetic resources on a global scale. This situation is due, in part, to the lack of a well-supported international institution that is specifically mandated to coordinate and facilitate the conservation and management of forest genetic resources. The FAO and the IUFRO are serving important functions in this regard, but a greater effort is needed. The pressures on forest genetic resources in industrialized and nonindustrialized countries differ in many respects, but the stability of future forest genetic resources for both could be enhanced through greater international leadership and coordination.

For the industrialized countries of the temperate and boreal regions, there are well-established species of widespread commercial use, populations of some of which have been widely sampled in parks, test stands, and in breeding and production stands. Many of those species are still incompletely sampled for conservation purposes, however, and for virtually all of them clear programs still do not exist for using genotypes or populations as introductions or substitutes for their current breeding population(s). The number of supporting populations for those species could profitably be at least doubled within existing national and international programs, and systems for testing and breeding enhanced alternatives could also be profitably established. In that regard, the number of species with clear potential for future use that could be managed in breeding or prebreeding operations is at least twice the number of species currently used, and their inclusion would quadruple the number of populations used. The flow of genetic materials from conserved status into advanced breeding populations would also have to be clearly established for this second set of species if they are to be used effectively. Hence, those species also may be justifiably conserved.

A third set of species, equal in number to the second set, could also be tested and conserved for future production forestry. The attention of the IUFRO and national agencies (with some interregional cooperation) is focused primarily on the first set, somewhat on the second, but little on the third. Largely independent interest groups also exist for nonproduction forestry species, which overlap with the third set of species. They focus on ecosystem conservation, however; very little research attention is given to the genetics of conservation or to ameliorative interventions.

In the nonindustrialized world, pressure is growing for production forestry, especially for fuelwood in dry areas but also for timber and other wood products. Breeding is largely just beginning (with a few exceptions, such as for *Eucalyptus*), and the structure of breeding populations has not yet been clearly defined. Hence, a need exists for founding a structure of primary and alternative breeding populations of all species that are prime candidates for production forestry, such as those that are being initiated by the CAMCORE, CSIRO, CTFT, FAO, and OFI. However, many of those activities are still focused on the development of test plantings for prime, potentially useful species, and that requires a large collection and testing array. Moreover, several hundred species of potential value, especially in the arid and the dry and moist tropical areas, are not included. Those species are at high risk of immediate loss, and both in situ and ex situ efforts by national governments, international cooperatives, and international donor programs are needed to protect them. At even greater risk, however, are those species of no obvious value to production forestry that are not managed or conserved and for which ecosystem conservation programs may or may not be structured to conserve any within-species diversity. There is no global program for systematically surveying those species for potential use, and no efforts are under way to develop the means for their restoration or use in afforestation.

RECOMMENDATIONS

The number of species with potential value that are included in testing and breeding programs should be at least doubled.
Many national and international efforts are directed toward conserving and managing forest tree species and populations. There is, however, a greater number of forest trees with potential value that should be included in conservation, testing, and breeding programs.

Sustained political support and expanded independent funding must be provided for long-term forest conservation operations, for professional and technical staff training, and for stabilization of institutions that address the needs of tree genetic resource conservation and management.
To increase forest tree resources in national and international programs for biological diversity conservation or resource development will require new and expanded programs and increased levels of funding. Long-term commitment of funds is necessary to assure the continuance of efforts that can extend over many years. Increased activity will create a need for more trained professional, technical, and support staff.

The capacity of national and regional institutions to manage tree genetic resources, especially in areas where loss of species and populations is most severe, should be enlarged and strengthened. Forest genetic resource conservation programs should be formalized and included in national plans for forestry, biological diversity, and breeding or other forms of genetic management.
With the current crisis in the availability of genetic material at all

these levels of use and management, a more secure global system is needed to ensure future access to the extant genetic resource. Nationally based organizations must take responsibility for any populations that originate entirely within the country's borders and for which the uses and benefits are also contained therein. They may require support and assistance to do so, but for their own local benefit. If the benefits and uses extend beyond national boundaries, however, then international interest in conserving and using those resources exists, and direct cooperative support is warranted. The benefits may be realized in terms of global health and ecosystem support, and more direct products and services, in which case the assistance of international agencies as well as cooperation may be needed to ensure that all appropriate technologies are used.

6

Organizing a Global System of Cooperation

National, regional, and international programs to conserve, manage, and use forest genetic resources are critical activities. Despite a wide variety of activities, the need remains for global leadership to coordinate and promote actions in an effective way. The overall goal should be to conserve and maintain forest genetic resources for both current and future needs, of which some may be presently unforeseen. There is a critical need for an organization with adequate and consistent funding and with the responsibility to provide scientific overview to set global priorities and to foster the necessary activities at national, regional, and global levels.

Multiple approaches must be used due to the significant gaps in the understanding of how to conserve forest genetic resources both in situ and ex situ, as well as the need to address immediately the high rate of loss for some species and regions. Coordination is needed to ensure excessive overlap in activities does not occur and important resources are not overlooked. No great simplification of the diverse collection of institutions and activities described here is envisioned. It is urgent, however, that key institutions be identified and that they be strengthened by charging them with specific responsibilities for conservation, management, or research and providing funds where needed to enable them to participate in a global network of activities.

COORDINATING AND EXPANDING EXISTING PROGRAMS

The achievement of global cooperation in managing tree genetic resources should begin with the coordination of existing efforts. Coordination would help to reduce costs and duplication of effort, and to increase sharing of both tree genetic resources and information. In particular, existing programs should be reviewed and where necessary strengthened. The coordination of these various efforts should seek to

• Develop and apply in situ and ex situ methods for conserving tree genetic diversity, recognizing that these approaches are complementary and supplement each other.

• Establish in situ reserves and ex situ conservation stands for a broad range of species because of the long generation times usually needed to obtain reproductively mature trees. Particular emphasis should be placed on species from tropical and arid regions.

• Promote research to develop and apply methods of ex situ storage to those species for which current technologies are not suitable and, in particular, for those species endangered in their natural environments. In addition, efforts are needed to foster the application of ex situ technologies for seed, pollen, and tissue storage, including cryogenic storage. In some cases, technologies are currently available and easily adopted. Adaptation and testing may be needed for other species or

The identification of tree genetic resources for use in agroforestry systems is gaining importance worldwide. Scientists at the Central Arid Research Institute, Jodhpur, India, conduct field trials with *Acacia* sp., *Prosopis* sp., and other woody species that may grow well in semiarid climates. Credit: Stanley Krugman.

methods. For still other techniques, such as cryogenic storage of tissues, further research and development will be necessary.

• Promote integration of conservation activities with testing and breeding programs.

• Assemble, analyze, and disseminate information about conservation efforts and programs for use as a base on which to build further activities in tree germplasm conservation.

A global system should strengthen the interaction among the various groups involved in managing forest genetic resources, coordinate actions, and facilitate the establishment of common priorities. Through a coordinating body, a global effort could foster and finance scientific research and training and monitor the status of conservation programs. The coordinating body could also cooperate with conservation programs of such groups as the International Union for the Conservation of Nature and Natural Resources and the United Nations Environment Program, which could incorporate activities specifically designed to monitor and conserve tree germplasm of global concern.

Research in a number of specific areas could enhance the knowledge base to support the conservation and management of forest genetic resources. The priority for these is likely to vary in different regions depending on local or regional needs, the information available, and funding or staff resources. Research is needed into the following:

• Reproductive biology and patterns of variation in priority species;

• Minimum population sizes necessary to maintain evolutionary flexibility of populations in nature reserves;

• Inventory of forest genetic resources at the regional and local level, particularly with respect to the distribution of natural populations;

• The size, design, and number of in situ reserves required to conserve a wide range of the genetic variation within a species, as well as the biological and ecological processes that affect species so conserved;

• Ex situ maintenance, evaluation, testing, and breeding technologies to enhance the preservation and use of conserved germplasm; and

• The effects of such processes as deforestation, acid rain, pollution, and global climatic change on the erosion of genetic variability and the integrity of in situ reserves and ex situ stands.

Within existing institutions there is a need for experienced, trained junior staff. Senior professional staff are also few in number. This severely constrains the ability of the institutions to expand their efforts. Because it is the experience of these trained and experienced staff that will form the basis for expanded programs, any new or enlarged programs must expect that it may take as long as 10 years to develop adequately prepared senior staff and proportionately less time to develop junior and field staff.

DEVELOPING REGIONAL AND NATIONAL PROGRAMS

Efforts must be made to develop comprehensive national capabilities for conserving forest tree genetic resources. Many countries already have organized units within forestry agencies or ministries to conduct various activities related to managing forests and forest genetic resources, including trees. It is within such programs that the management of forest genetic resources should occur. In many cases, for example, reserve and protected areas set aside as part of efforts to conserve pristine vegetation could also serve as in situ conservation stands for tree genetic resources. Forestry bureaus, departments, agencies, or ministries often operate in isolation, however, and do not interact with those concerned with natural resources or agriculture. Frequently their principal function is regulatory rather than research and conservation oriented.

Comprehensive, coordinated national programs are needed to

• Develop national policies for the genetic conservation of trees and establish priority activities to accomplish that objective;

• Monitor the status of tree species that are of economic, ecological, or aesthetic importance;

• Develop programs for implementing in situ and ex situ management and conservation of tree germplasm to ensure that the genetic diversity that exists within and among tree species is not lost;

• Develop programs to promote the flow of conserved materials from storage programs to testing and breeding programs;

• Promote and support training and research for persons working in tree germplasm conservation at universities, colleges, experimental stations, and institutes and in the private sector; and

• Foster and support programs to maintain data related to forest genetic resources activities and promote the evaluation and use of tree germplasm to meet national needs.

The first actions to institute national programs will likely be taken by those agencies or ministries concerned with forest resources. Many activities related to forest conservation are already in place and could support the management of tree genetic resources. Further, research applicable to tree germplasm management can most easily be developed within universities or existing national forestry institutions. Additionally, a number of activities could be conducted in cooperation with other

Many reforestation efforts undertaken by governments to halt erosion have been in place for many years. This hillside near Lumaco, Chile, was part of a reforestation effort in 1971 in which the government, through the Corporation for Agrarian Reform, planted 68 million tree seedlings including *Pinus radiata*, a species that flourishes under poor conditions. Credit: Food and Agriculture Organization.

national conservation programs. Long-term seed storage, for example, might be handled jointly with a national agricultural gene bank. In situ conservation areas, for another example, could be coordinated with a national park system. The exchange and introduction of tree germplasm could be the cooperative efforts of forestry and agricultural interests.

Regional programs are uniquely suited to addressing the needs of and facilitating activities among national programs that have common interests and priorities. Regional institutions can assist member countries in developing national and cooperative activities related to forest genetic resource conservation. They may provide funds for programs to inventory genetic resources, collect species, and establish nurseries and ex situ stands and to promote training, research, and scientific exchange within and outside the region. Programs for forestry genetic resources may be organized within the framework of a regional institution, as for example, that of Centro Agronómico Tropical de Investigación y Enseñanza in Costa Rica. Many activities of regional programs are similar to those of global programs. They are, however, focused more on the unique species, needs, and constraints of the region. Further, because the occurrence of trees is seldom restricted to a single nation, regional programs are better able than national programs to coordinate efforts over the entire range of a species.

Regional programs should be developed to

• Assess the regional status of forest genetic resources and develop priorities for their conservation;

• Facilitate information and germplasm exchanges within the region and act as liaison to programs outside the region;

• Develop training programs that focus on the species and technical needs important to the region; and

• Foster cooperative research programs to address regionally specific problems by supplying funds and sponsoring meetings, workshops, or symposia.

Regional organizations should facilitate interactions among national programs and promote or enhance cooperation with international groups and programs. By coordinating the activities of national programs, regional programs could enable work to proceed on a broad range of species in a cost-effective and efficient manner. Governments or programs with limited resources would be able to focus on their highest priority needs, while being assured that species that are of concern, but for which resources may be scarce, are being addressed elsewhere in the region.

DEVELOPING AN INTERNATIONAL FOREST TREE GENETIC RESOURCES PROGRAM

A global institution charged with addressing global concerns for managing forest tree genetic resources would maintain an ongoing assessment of the status of forest genetic conservation worldwide and foster the study, collection, documentation, evaluation, and use of tree genetic resources. It would facilitate interaction among regional and national programs, support and encourage training at all levels, support research and its application to managing forest tree germplasm, act as a central source for assembling and disseminating data to national and regional programs, and where necessary, provide funding for conservation activities it identifies as having high priority. Specifically, an international body could

• Establish global priorities, by species, for genetic resource conservation and assess the extent to which those priorities are addressed by existing programs;

Cattle graze on this ranch in Brazil, which was created by clearing and burning a portion of the rain forest. Credit: James P. Blair ©National Geographic Society.

- Encourage and assist national and regional programs in planning, organizing, and managing forest genetic resources programs;
- Encourage and support training and research by national and regional programs;
- Assemble and disseminate data relevant to managing gene banks and in situ reserves; and
- Facilitate interaction among programs to encourage the exchange of information, technology, and germplasm through meetings, conferences, and symposia.

Such an institution could most easily be developed as part of an existing international agency, or come from the expansion or modification of an existing program. This global body should function autonomously under a board, institutional program, or committee structure. It should seek to develop relationships with other international conservation activities to promote cooperation and collaboration in areas of mutual interest.

It is imperative that the leadership of a global institution for conserving tree genetic resources be recognized by participating countries, institutions, and programs. The enormity of such a task and its permanent nature require the participation of all governments as well as many private and public institutions. Several organizations could conceivably assume the global function of providing cohesiveness and direction to national, regional, and international efforts to conserve and use forest genetic resources.

A global effort must be involved not only with seed and other ex situ storage, but also with in situ conservation and in networks of managed observation, test, and conservation stands. Coordination and continuous management of relationships among many national agencies are necessary. Thus, the most effective program might involve cooperation of two or more institutions with appropriate expertise.

The committee has identified two organizations, the Food and Agriculture Organization (FAO) of the United Nations and the International Board for Plant Genetic Resources (IBPGR), that could serve this purpose. Other existing institutions might also participate. Discussions within the Consultative Group for International Agricultural Research (CGIAR) may lead to development of a global forestry program within that organization.

Coordination Through the Food and Agriculture Organization

The FAO could be a focal point for international planning and coordination based on its historical role in developing global forest management programs. It would, however, need substantially greater direct support to take on the responsibilities outlined herein. The agencies that are regionally active also would have to expand their testing and breeding programs. The FAO, independently or in cooperation with the International Union of Forestry Research Organizations, could promote scientific evaluations of the adequacy of all programs with the aim of enhancing those activities and fostering or encouraging new ones. All of this would require considerable increases in the FAO's funding and its technical staff.

Coordination Through the International Board for Plant Genetic Resources

At its organization in the early 1970s as an international agricultural research center, the IBPGR was expected to include attention to tree species in its activities. Lack of sufficient funds prevented this, however, and currently it does not deal with forest species. The IBPGR's mandate would need little change to incorporate forest tree resources in its program. It would have to develop an institutional structure, however,

At an elevation of about 2,000 m in the Cameron highlands of Malaysia, villagers grow vegetables where tropical forests once grew. National and international policies and programs must balance conservation concerns with the needs of growing populations. Credit: James P. Blair ©National Geographic Society.

to ensure that forest tree resources issues were addressed by those with appropriate professional expertise. The IBPGR's scientific and technical capabilities are largely in the area of agricultural genetic resources. Thus, it would have to assemble additional professional expertise, restructure its board, and obtain the additional funds necessary to take on these added responsibilities. The existing technologies and resources available for crop genetic resources could make resource use for a forest genetic resources program more efficient and could greatly facilitate the implementation of such a program.

RECOMMENDATION

An international institution should be established or designated to provide leadership, coordination, and facilitation for the global management of the world's tree genetic resources at national, regional, and international levels.

There is an obvious need to expand programs in genetic resource development for production in tropical- as well as temperate-zone forestry. A related need also exists for vast expansion of research and development programs for genetic conservation and development of nonproduction forests. No agency, however, has the financial or personnel resources capable of organizing or managing such expansions. In fact, the number of trained personnel available for such expanded programs is inadequate, and scientific and managerial support is in critically short supply, especially in developing nations.

A multifaceted, long-term global program must urgently be developed to manage the genetic resources of the world's forests. In view of current constraints, the committee considers a sharing of responsibilities by the FAO and a forestry body within the IBPGR (possibly as part of a larger CGIAR effort in forestry), augmented by expertise from other international, regional, and national programs, to be the most feasible approach.

References

Ashton, P. S. 1969. Speciation among tropical forest trees: Some deductions in the light of recent evidence. Biol. J. Linn. Soc. 1:155–196.

Ashton, P. S. 1981. Techniques for the identification and conservation of threatened species in tropical forests. Pp. 155–164 in The Biological Aspects of Rare Plant Conservation, H. Synge, ed. New York: John Wiley & Sons.

Ashton, P. S. 1988. Dipterocarp reproductive biology. Pp. 219–240 in Ecosystems of the World. Vol. 14B, Tropical Rain Forest Ecosystems: Biogeographical and Ecological Studies, H. Lieth and M. J. A. Verger, eds. Amsterdam: Elsevier.

Baum, W. C. 1986. Partners Against Hunger: The Consultative Group on International Agricultural Research. Washington, D.C.: World Bank.

Bawa, K. S. 1974. Breeding systems of tree species of a lowland tropical community. Evolution 28:85–92.

Bawa, K. S. 1976. Breeding of tropical hardwoods: An evaluation of underlying bases, current status, and future prospects. Pp. 43–59 in Tropical Trees: Variation, Breeding, and Conservation, J. Burley and B. T. Styles, eds. London: Academic Press.

Bawa, K. S. 1979. Breeding systems of trees in a tropical wet forest. N. Z. J. Bot. 17:521–524.

Bawa, K. S., and P. A. Opler. 1975. Dioecism in tropical forest trees. Evolution 29:167–179.

Bawa, K. S., D. R. Perry, and J. H. Beach. 1985. Reproductive biology of tropical lowland rain forest trees. I. Sexual systems and incompatibility mechanisms. Amer. J. Bot. 72:331–345.

Bergman, F., and H.-R. Gregorius. 1979. Comparison of the genetic diversity of various populations of Norway spruce (Picea abies). Pp. 99–107 in Proceedings of the Conference on the Biochemical Genetics of Forest Trees. Umea, Sweden: Swedish University of Agricultural Sciences.

Bonner, F. T. 1985. Technologies to Maintain Tree Germplasm Diversity. Prepared for the Office of Technology Assessment, U.S. Congress, Washington, D.C.

Bonner, F. T. 1990. Storage of seeds: Potential and limitations for germplasm conservation. In Proceedings of the Symposium on the Conservation of Genetic Diversity, C. Millar and F. T. Ledig, eds. For. Ecol. Manage. 35:35–43.

139

Brotschol, J. V. 1983. Allozyme Variation in Natural Populations of *Liriodendron tulipifera* L. Ph.D. dissertation, North Carolina State University, Raleigh.

Brotschol, J. V., J. H. Roberds, and G. Namkoong. 1986. Allozyme variation among North Carolina populations of *Liriodendron tulipifera* L. Silvae Genet. 35:131–138.

Burley, J. 1987. Exploitation of the potential of multipurpose trees and shrubs in agroforestry. In Agroforestry—A Decade of Development, H. Steppler and P. K. R. Nair, eds. Nairobi, Kenya: International Council for Research in Agroforestry.

Burley, J., and P. J. Wood. 1976. A Manual on Species and Provenance Research with Particular Reference to the Tropics. Oxford: Commonwealth Forestry Institute.

Burley, J., and P. J. Wood. 1987. A Tree for All Reasons; Evaluation of Multipurpose Trees. Nairobi: International Council for Research in Agroforestry.

Burley, J., and P. von Carlowitz, eds. 1984. Multipurpose Tree Germplasm. Proceedings of a Planning Workshop to Discuss International Cooperation. Nairobi: International Council for Research in Agroforestry.

Consultative Group on International Agricultural Research (CGIAR). 1989a. Proposed expansion of the CGIAR. ICW/89/09 Agenda Item 8(c). Prepared for International Centers Week, October 30 to November 3, 1989, Washington, D.C.

CGIAR. 1989b. A possible expansion of the CGIAR: Part I interim report. AGR/TAC:IAR/88/24 Add.2 Agenda Item 8(a). Prepared for International Centers Week, October 30 to November 3, 1989, Washington, D.C.

CGIAR. 1990. International Centers Week 1990, Summary of Proceedings and Decisions. Washington, D.C.: CGIAR Secretariat.

Council on Environmental Quality and U.S. Department of State. 1980. The Global 2000 Report. Vol. 2, Technical Report. Washington, D.C.: U.S. Government Printing Office.

Cowell, A. 1990. Decade of Destruction. New York: Henry Holt.

Croat, T. B. 1978. Flora of Barra Colorado Island. Palo Alto, Calif.: Stanford University Press.

El-Kassaby, Y. A. 1982. Associations between allozyme genotypes and quantitative traits in Douglas-fir. Genetics 101:103–115.

Eriksson, G., I. Ekbert, and A. Jonsson. 1972. Meiotic and pollen investigations as a guide for localization of forest tree seed orchards in Sweden. Pp. 1–28 in Proceedings of the Joint Symposia for Forest Tree Breeding of Genetics Subject Group. Tokyo: Government Forestry Experiment Station.

Evans, J. 1982. Plantation Forestry in the Tropics. New York: Oxford University Press.

Ewens, W. J., P. J. Brockwell, J. M. Gani, and S. I. Resnick. 1987. Minimum viable population size in the presence of catastrophes. Pp. 59–68 in Viable Populations for Conservation, M. Soulé, ed. New York: Cambridge University Press.

Falkenhagen, E. R. 1985. Isozyme studies in provenance research of forest trees. Theor. Appl. Genet. 69:335–347.

Fins, L., and L. W. Seeb. 1986. Genetic variations in allozymes of western larch. Can. J. For. Res. 16:1013–1018.

Food and Agriculture Organization (FAO). 1969. Report of the First Session of the FAO Panel of Experts on Forest Gene Resources. FO:FGR/1/Rep. Rome, Italy: Food and Agriculture Organization.

FAO. 1972. Report of the Second Session of the FAO Panel of Experts on Forest Gene Resources. FO:FGR/2/Rep. Rome, Italy: Food and Agriculture Organization.

FAO. 1974. Report of the Third Session of the FAO Panel of Experts on Forest Gene Resources. FO:FGR/3/Rep. Rome, Italy: Food and Agriculture Organization.

FAO. 1977. Report of the Fourth Session of the FAO Panel of Experts on Forest Gene Resources. FO:FGR/4/Rep. Rome, Italy: Food and Agriculture Organization.

FAO. 1981. Forest Genetic Resources Information, No. 10. Rome, Italy: Food and Agriculture Organization.

FAO. 1984a. In-Situ Conservation of Genetic Resources of Plants: The Scientific and Technical Base. FORGEN/MISC/84/1. Rome, Italy: Food and Agriculture Organization.

FAO. 1984b. A Guide to In-Situ Conservation of Genetic Resources for Tropical Woody Species. FORGEN/MISC/84/2. Rome, Italy: Food and Agriculture Organization.

FAO. 1984c. In-Situ Conservation of Wild Plant Genetic Resources: A Status Review and Action Plan. FORGEN/MISC/84/3. Rome, Italy: Food and Agriculture Organization.

FAO. 1985a. Report of the Sixth Session of the FAO Panel of Experts on Forest Gene Resources. FO:FGR/5/Rep. Rome, Italy: Food and Agriculture Organization.

FAO. 1985b. A World Perspective; Forestry Beyond 2000. Unasylva 147:7–16.

FAO. 1985c. The Tropical Forest Action Plan. Prepared by the FAO in cooperation with the World Resources Institute, World Bank, and United Nations Development Program. Rome, Italy: Food and Agriculture Organization.

FAO. 1985d. In-situ conservation of forest genetic resources in Cameroon. Pp. 15–30 in Forest Genetic Resources Information, No. 14. Rome, Italy: Food and Agriculture Organization.

FAO. 1985e. In-situ conservation of forest genetic resources in peninsular Malaysia. Pp. 32–49 in Forest Genetic Resources Information, No. 14. Rome, Italy: Food and Agriculture Organization.

FAO. 1986. Databook on Endangered Tree and Shrub Species and Provenances. FAO Forestry Paper 77. Rome, Italy: Food and Agriculture Organization.

FAO. 1987. In-situ conservation in Peru: A case study. Pp. 5–21 in Forest Genetic Resources Information, No. 15. Rome, Italy: Food and Agriculture Organization.

FAO. 1988. Production Yearbook, 1987. FAO Statistics Series No. 82. Rome, Italy: Food and Agriculture Organization.

Food and Agriculture Organization (FAO) and United Nations Environment Program (UNEP). 1981. Report on the FAO/UNEP Expert Consultation on In-Situ Conservation of Forest Genetic Resources, December 2–4, 1980. Rome, Italy: Food and Agriculture Organization.

Fowler, D. P., and R. W. Morris. 1977. Genetic diversity in red pine: Evidence for low genic heterozygosity. Can. J. For. Res. 7:343–347.

Frankel, O. H., and M. E. Soulé. 1981. Conservation and Evolution. New York: Cambridge University Press.

Frankie, G. W., P. A. Opler, and K. S. Bawa. 1976. Foraging behavior of solitary bees: Implications for outcrossing of a neotropical forest tree species. J. Ecol. 64:1049–1057.

Franklin, I. R. 1980. Evolutionary change in small populations. Pp. 135–149 in Conservation Biology: An Evolutionary-Ecological Perspective, M. E. Soulé and B. A. Wilcox, eds. Sunderland, Mass.: Sinauer Associates.

Gentry, A. 1986. Endemism in tropical versus temperate plant communities. Pp. 153–181 in Conservation Biology: The Science of Scarcity and Diversity, M. E. Soulé, ed. Sunderland, Mass.: Sinauer Associates.

Gilbert, L. E. 1980. Food web organization and conservation of neotropical diversity. Pp. 11–54 in Conservation Biology: An Evolutionary-Ecological Perspective, M. E. Soulé and B. A. Wilcox, eds. Sunderland, Mass.: Sinauer Associates.

Gilpin, M. E., and M. E. Soulé. 1986. Minimum viable populations: Processes of species extinction. Pp. 19–34 in Conservation Biology: The Science of Scarcity and Diversity, M. E. Soulé, ed. Sunderland, Mass.: Sinauer Associates.

Govindaraju, D. R. 1988a. A note on the relationship between outcrossing rate and gene flow in plants. Heredity 61:401–404.

Govindaraju, D. R. 1988b. Relationship between dispersal ability and levels of gene flow in plants. Oikos 52:31–35.

Grainger, A. 1987. The future of the tropical moist forest. Paper prepared for Resources for the Future, Washington, D.C.

Gregorius, H.-R. 1984. A unique genetic distance. Biom. J. 26:13–18.

Gregorius, H.-R., and G. Namkoong. 1983. Conditions for protective polymorphism in subdivided plant populations. Theor. Pop. Biol. 24:252–267.

Gregorius, H.-R., and J. H. Roberds. 1986. Measurement of genetical differentiation among subpopulations. Theor. Appl. Genet. 71:826–834.

Hamrick, J. L. 1983. The distribution of genetic variation within and among natural plant populations. Pp. 335–348 in Genetics and Conservation, C. M. Schonewald-Cox, S. M. Chambers, B. MacBryde, and W. L. Thomas, eds. Menlo Park, Calif.: Benjamin/ Cummings.

Hamrick, J. L., and M. D. Loveless. 1987. The influence of seed dispersal mechanisms on the genetic structure of plant populations. Pp. 211–223 in Frugivores and Seed Dispersal, A. Estrada and T. H. Fleming, eds. The Hague: W. Junk.

Hamrick, J. L., Y. B. Linhart, and J. B. Mitton. 1979. Relationships between life history characteristics and electrophoretically detectable genetic variation in plants. Annu. Rev. Ecol. Syst. 10:173–200.

Harlan, H. V., and M. L. Martini. 1936. Problems and results in barley breeding. Pp. 303–346 in USDA Yearbook of Agriculture. Washington, D.C.: U.S. Department of Agriculture.

Hegde, N. G., and P. D. Abhyankar, eds. 1986. The Greening of Wastelands. Pune, India: Bharatiya Agro Industries Foundation.

Heithaus, E. R., T. H. Fleming, and P. A. Opler. 1975. Foraging patterns and resource utilization in seven species of bats in a seasonal tropical forest. Ecology 56:841–854.

Hiebert, R. D., and J. L. Hamrick. 1983. Patterns and levels of genetic variation in Great Basin bristlecone pine, Pinus longaeva. Ecology 37:302–310.

Hubbell, S. P., and R. B. Foster. 1986. Commonness and rarity in a neotropical forest: Implications for tropical tree conservation. Pp. 205–231 in Conservation Biology: The Science of Scarcity and Diversity, M. E. Soulé, ed. Sunderland, Mass.: Sinauer Associates.

International Board for Plant Genetic Resources. 1989. Annual Report 1988. Rome, Italy: International Board for Plant Genetic Resources.

International Council for Research in Agroforestry. 1990. Annual Report 1989. Nairobi, Kenya: International Council for Research in Agroforestry.

International Union for the Conservation of Nature and Natural Resources. 1978. Categories, Objectives, and Criteria for Protected Areas. Gland, Switzerland: International Union for the Conservation of Nature and Natural Resources.

International Union for the Conservation of Nature and Natural Resources, United Nations Environment Program, and World Wildlife Fund. 1980. World Conservation Strategy: Living Resource Conservation for Sustainable Development. Gland, Switzerland: International Union for the Conservation of Nature and Natural Resources.

International Union of Forestry Research Organizations (IUFRO). 1985. Increasing Productivity of Multipurpose Species, J. Burley and J. L. Stewart, eds. Report on a Planning Workshop for Asia on Forest Research and Technology Transfer. Vienna, Austria: International Union of Forestry Research Organizations.

IUFRO. 1986. Increasing Productivity of Multipurpose Lands, L. W. Carlson and K. R. Shea, eds. Research Planning Workshop for Africa: Sahelian and North Sudanian Zones. Vienna, Austria: International Union of Forestry Research Organizations.

IUFRO. 1987. Tree Improvement and Silvo-Pastoral Management in Sahelian and North Sudanian Africa; Problems, Needs, and Research Proposals, C. Cossalter, D. E. Iyambo, S. L. Krugman, and O. Fugalli, compilers. Vienna, Austria: International Union of Forestry Research Organizations.

Janzen, D. H. 1971. Euglossine bees as long distance pollinators of tropical plants. Science 171:203–205.

Janzen, D. H. 1979. How to be a fig. Annu. Rev. Ecol. Syst. 10:13–51.

Janzen, D. H. 1987. Insect diversity of a Costa Rican dry forest: Why keep it, and how? Biol. J. Linn. Soc. 30:343–356.

King, K. F. S., and T. Chandler. 1978. Wasted Lands. Nairobi: International Center for Research in Agroforestry.

Kloppenburg, Jr., J. R., ed. 1988. Seeds and Sovereignty: The Use and Control of Plant Genetic Resources. Durham, N.C.: Duke University Press.

Knowles, P., G. R. Furnier, M. A. Aleksiuk, and D. J. Perry. 1987. Significant levels of self-fertilization in natural populations of tamarack. Can. J. Bot. 65:1087–1091.

Landauer, W. 1945. Shall we lose or keep our plant and animal stocks? Science 101:497–499.

Lanly, J. P. 1982. Tropical Forests Resources. FAO Forestry Paper 30. Rome, Italy: Food and Agriculture Organization.

Ledig, F. T. 1986. Heterozygosity, heterosis and fitness in outbreeding plants. Pp. 77–104 in Conservation Biology: The Science of Scarcity and Diversity, M. E. Soulé, ed. Sunderland, Mass.: Sinauer Associates.

Ledig, F. T. 1987. Genetic structure and the conservation of California's endemic and near-endemic conifers. Pp. 587–594 in Conservation and Management of Rare and Endangered Plants, T. S. Ellis, ed. Sacramento: California Native Plant Society.

Ledig, F. T., and M. T. Conkle. 1983. Gene diversity and genetic structure in a narrow endemic, Torrey pine (*Pinus torreyana* Parry ex Carr). Evolution 37:79–85.

Levin, D. A., and H. W. Kerster. 1974. Gene flow in seed plants. Evol. Biol. 7:139–220.

Libby, W. J., and W. B. Critchfield. 1987. Patterns of genetic architecture. Ann. For. 13:77–92.

Linhart, Y. B. 1973. Ecological and behavioral determinants of pollen dispersal in hummingbird pollinated *Heliconia*. Amer. Nat. 107:511–523.

Linhart, Y. B., J. B. Mitton, K. B. Sturgeon, and M. L. Davis. 1981a. An analysis of genetic architecture in populations of ponderosa pine. Pp. 53–59 in Proceedings of the Symposium on Isozymes of North American Forest Trees and Forest Insects. General Technical Report PSW-48. Berkeley, Calif.: U.S. Forest Service, Pacific Southwest Forest and Range Experiment Station.

Linhart, Y. B., J. B. Mitton, K. B. Sturgeon, and M. L. Davis. 1981b. Genetic variation in space and time in a population of ponderosa pine. Heredity 46:407–426.

Loveless, M. D., and J. L. Hamrick. 1984. Ecological determinants of genetic structure in plant populations. Annu. Rev. Ecol. Syst. 15:65–95.

Marshall, A. G. 1983. Bats, flowers and fruit: Evolutionary relationships in the Old World. Biol. J. Linn. Soc. 20:115–135.

Mergen, F., and J. R. Vincent, eds. 1987. Natural Management of Tropical Moist Forests. New Haven, Conn.: Yale University School of Forestry and Environmental Studies.

Moran, G. F., and S. D. Hopper. 1987. Conservation of genetic resources of rare and widespread eucalypts in remnant vegetation. Pp. 151–162 in Nature Conservation: The Role of Remnants of Native Vegetation, D. A. Saunders, G. W. Arnold, A. A. Burbidge, and A. J. M. Hopkins, eds. Chipping Norton, NSW, Australia: Surrey Beatty & Sons.

Moran, G. F., and W. T. Adams. 1989. Microgeographical patterns in allozyme differentiation in Douglas-fir from southwest Oregon. For. Sci. 35:3–15.

Myers, N. 1983a. Tropical moist forests: Over-exploited and under-utilized? For. Ecol. Manage. 6(1):59–79.

Myers, N. 1983b. A Wealth of Wild Species. Boulder, Colo.: Westview Press.

Myers, N. 1988. Tropical forests: Much more than stocks of wood. J. Trop. Ecol. 4:209–221.

Namkoong, G. 1969. The non-optimality of local races. Pp 149–153 in Proceedings of the 10th Southern Conference on Forest Tree Breeding, Texas Forest Service, Texas A&M University. College Station, Tex.: University Press.

Namkoong, G. 1984. Genetic structure of forest tree populations. Pp. 351–360 in Genetics: New Frontiers, V. L. Chopra, B. C. Joshi, R. P. Sharma, and H. C. Bansal, eds. Vol. 4, Applied Genetics. Proceedings of the 15th International Congress of Genetics. New Delhi: Mohan Primlani, Oxford, and IBH Publishers.

Namkoong, G., and H.-R. Gregorius. 1985. Conditions for protected polymorphism in subdivided plant populations. 2. Seed versus pollen migration. Amer. Nat. 125:521–534.

Namkoong, G., and H. C. Kang. 1990. Quantitative genetics of forest trees. Pp. 139–188 in Plant Breeding Reviews, Vol. 8. Portland, Ore.: Timber Press.

Namkoong, G., R. D. Barnes, and J. Burley. 1980. A Philosophy of Breeding Strategy for Tropical Forest Trees. Tropical Forestry Paper No. 16. Oxford: Commonwealth Forestry Institute.

National Research Council. 1990. Forestry Research: A Mandate for Change. Washington, D.C.: National Academy Press.

Nei, M. 1972. Genetic distance between populations. Amer. Nat. 106:283–292.

Office of Technology Assessment. 1984. Technologies to Sustain Tropical Forest Resources. OTA-F214. Washington, D.C.: U.S. Government Printing Office.

Oldfield, M. L. 1984. The Value of Conserving Genetic Resources. Washington, D.C.: National Park Service, U.S. Department of the Interior.

O'Malley, D. M., and K. S. Bawa. 1987. Mating system of a tropical rain forest tree. Amer. J. Bot. 74:1143–1149.

Palmberg, C. 1988. Plant Genetic Resources: Their Conservation In Situ for Human Use. Rome, Italy: Food and Agriculture Organization.

Palmberg, C., and J. T. Esquinas-Alcázar. 1990. The role of international organizations in the conservation of plant genetic resources, with special reference to forestry and the UN system. Proceedings of the Symposium on the Conservation of Genetic Diversity. For. Ecol. Manage. 35:171–197.

Panday, K. 1982. Fodder Trees and Tree Fodder in Nepal. Berne and Birmensdorf: Swiss Development Corporation and Swiss Federal Institute of Forestry Research.

Parkash, R., and D. Hocking. 1986. Some Favourite Trees for Fuel and Fodder. New Delhi: Society for Promotion of Wastelands Development.

Pearce, F. 1990. Brazil, where the ice cream comes from. New Scientist, July 7, 45–48.

Peters, W. J., and L. N. Neuenschwander. 1988. Slash and Burn Farming in the Third World. Moscow: University of Idaho Press.

Pimm, S. L. 1986. Community stability and structure. Pp. 309–332 in Conservation Biology: The Science of Scarcity and Diversity, M. E. Soulé, ed. Sunderland, Mass.: Sinauer Associates.

Plucknett, D. L., N. J. H. Smith, and S. Ozgediz. 1990. International Agricultural Research: A Database of Networks. Washington, D.C.: World Bank.

Postel, S. 1989. Halting land degradation. Pp. 21–40 in State of the World 1989, L. Starke, ed. New York: W. W. Norton.

Postel, S., and L. Heise. 1988. Reforesting the Earth. Worldwatch Paper 83. Washington, D.C.: Worldwatch Institute.

Prance, G. T. 1984. Completing the inventory. Pp. 365–396 in Current Concepts in Plant Taxonomy, V. H. Heywood and D. M. Moore, eds. London: Academic Press.

Prescott-Allen, C., and R. Prescott-Allen. 1986. The First Resource. Wild Species in the North American Economy. New Haven, Conn.: Yale University Press.

Prescott-Allen, R., and C. Prescott-Allen. 1983. Genes from the Wild. Washington, D.C.: International Institute for Environment and Development.

Regal, P. J. 1977. Ecology and evolution of flowering plant dominance. Science 196:622–629.

Roberds, J. H., and M. T. Conkle. 1984. Genetic structure in loblolly pine stands: Allozyme variation in parents and progeny. For. Sci. 30:319–329.

Roche, L. R., ed. 1975. Methodology of Conservation of Forest Genetic Resources. Rome, Italy: Food and Agriculture Organization.

Ross, R. K. 1988. Patterns of Allelic Variation in Natural Populations of *Abies fraseri*. Ph.D. dissertation, North Carolina State University, Raleigh.

Sakei, K., and Y. Park. 1971. Genetic studies in natural populations of forest tress. Theor. Appl. Genet. 41:13–17.

Schoenwald-Cox, C. M., S. N. Chambers, B. MacBryde, and L. Thomas. 1983. Genetics and Conservation. Menlo Park, Calif.: Benjamin/Cummings.

Scholz, F., H.-R. Gregorius, and D. Rudin, eds. 1989. Genetic Effects of Air Pollutants in Forest Tree Populations. New York: Springer-Verlag.

Silver, C. S. 1990. One Earth, One Future: Our Changing Global Environment. Washington, D.C.: National Academy Press.

Simberloff, D. 1986. Are we on the verge of a mass extinction in tropical rain forests? Pp. 165–180 in Dynamics of Extinction, E. K. Elliot, ed. New York: John Wiley & Sons.

Slatkin, M. 1987. Gene flow and the geographic structure of natural populations. Science 236:787–792.

Soulé, M. E. 1980. Thresholds for survival: Maintaining fitness and evolutionary potential. Pp. 153–169 in Conservation Biology: An Evolutionary-Ecological Perspective, M. E. Soulé and B. A. Wilcox, eds. Sunderland, Mass.: Sinauer Associates.

Soulé, M. E., ed. 1986. Conservation Biology: The Science of Scarcity and Diversity. Sunderland, Mass.: Sinauer Associates.

Soulé, M. E., ed. 1987. Viable Populations for Conservation. New York: Cambridge University Press.

Steinhoff, R. J., D. G. Joyce, and L. Fins. 1983. Isozyme variation in *Pinus monticola*. Can. J. For. Res. 13:1122–1132.

Terborgh, J. 1986. Keystone plant resources in tropical forests. Pp. 330–344 in Conservation Biology: The Science of Scarcity and Diversity, M. E. Soulé, ed. Sunderland, Mass.: Sinauer Associates.

Terborgh, J. 1988. The big things that run the world—A sequel to E. O. Wilson. Conserv. Biol. 2:402–403.

Tigerstedt, P. M. A. 1984. Genetic mechanisms for adaptation: The mating system of Scots pine. Pp. 317–324 in Genetics: New Frontiers, V. L. Chopra, B. C. Joshi, R. P. Sharma, and H. C. Bansal, eds. Vol. 4, Applied Genetics. Proceedings of the 15th International Congress of Genetics. New Delhi: Mohan Primlani, Oxford, and IBH Publishers.

United Nations. 1984. International Tropical Timber Agreement, 1983. TD/TIMBER/11/Rev. 1. New York: United Nations.

U.S. Forest Service. 1982. 1981 Directory of Forest Tree Seed Orchards in the United States. FS-278. Washington, D.C.: U.S. Forest Service.

van Sloten, D. H. 1990. Facing the future—IBPGR in the 1990s. Paper prepared for International Centers Week, October 29–November 2, 1990, Washington, D.C., sponsored by the Consultative Group on International Agricultural Research.

von Carlowitz, L. C. 1986. Multipurpose Tree and Shrub Directory. Nairobi, Kenya: International Council for Research and Agroforestry.

Wellendorf, H., and A. Kaosa-Ard. 1988. Teak Improvement Strategy in Thailand. Arboretet Horsholm Forest Tree Improvement 21. Copenhagen: DSR Forlag.

Wiebes, J. T. 1979. Coevolution of figs and their insect pollinators. Annu. Rev. Ecol. Syst. 10:1–12.

Wilson, E. O. 1988. Biodiversity. Washington, D.C.: National Academy Press.

Withers, L. A. 1989. In vitro conservation and germplasm utilization. Pp. 309–334 in The Use of Plant Genetic Resources, A. D. H. Brown, O. H. Frankel, D. R. Marshall, and J. T. Williams, eds. Cambridge: Cambridge University Press.

Withers, L. A., and J. T. Williams, eds. 1982. Crop Genetic Resources—The Conservation of Difficult Material. Series B42. Paris: International Union of Biological Sciences.

Wood, P. J., J. Burley, and A. Grainger. 1982. Technologies and Technology Systems for Reforestation of Degraded Tropical Land. Prepared for the Office of Technology Assessment, U.S. Congress, Washington, D.C.

World Resources Institute, International Institute for Environment and Development, and United Nations Environment Program. 1988. World Resources, 1988–89. New York: Basic Books.

Wright, J. W. 1962. Genetics of Forest Tree Improvement. FAO Forestry and Forest Products Studies, No. 16. Rome, Italy: Food and Agriculture Organization.

Wyatt-Smith, J. 1987. The management of tropical moist forest for the sustained production of timber: Some issues. Tropical Forestry Policy Paper No. 4. Prepared for the International Union for the Conservation of Nature and Natural Resources, Gland, Switzerland, and the International Institute for Environment and Development, London, England.

Yeh, F. C., and C. Layton. 1979. The organization of genetic variability in central and marginal populations of lodgepole pine *Pinus contorta* spp. *latifolia*. Can. J. Genet. Cytol. 21:487–503.

Zobel, B., and J. T. Talbert. 1986. Applied Forest Tree Improvement. New York: John Wiley & Sons.

A

Forest Tree Species Used in Breeding or Testing Activities

The breeding and testing activities listed in Table A-1 reflect programs located in more than 90 countries. It was compiled from publications and from data supplied directly by universities, state and national forest services, and private companies.

There are 400 species listed as being involved in breeding or testing activities. The 176 species identified in Table A-1 as being involved in breeding (B) are those for which selection and mating have resulted in at least one breeding generation for at least one identified population. Breeding efforts with at least 20 seed parents number 35 and are denoted by B+. The 340 species identified as being involved in testing (T) are those that have been included in comparison tests of several species in replicated designs. Tests conducted in at least three different locations number 95 and are denoted by T+. Species listed as being involved in both breeding and testing, whether as independent or combined activities, number 116.

TABLE A-1 Forest Tree Species Included in at Least One Breeding or Testing Activity

Genus/Species	Breeding	Testing
Abies (fir)		
A. alba	B+	T
A. amabilis		T
A. balsamea	B	
A. bornmuelleriana		T
A. cephalonica	B	T
		(*continued*)

TABLE A-1 (Continued)

Genus/Species	Breeding	Testing
A. concolor	B	T
A. equitrojani		T
A. fraseri	B	
A. grandis	B	T+
A. holophylla	B	
A. magnifica var. shastensis		T
A. nordmanniana	B	T
A. procera		T+
A. sachalinensis		T
A. webbiana		T
Acacia (acacia)		
A. albida		T+
A. aneura		T+
A. arabica	B	
A. catechu	B	
A. caven		T
A. cowleana		T+
A. dealbata		T
A. decurrens		T
A. erioloba		T
A. farnesiana		T
A. holosericea		T+
A. mangium	B	T
A. mearnsii		T
A. melanoxylon		T
A. nilotica		T+
A. nilotica var. adansonii		T
A. nilotica var. tomentosa		T+
A. nilotica ssp. indica		T
A. nilotica ssp. indica var. cupressiformes		T
A. nilotica ssp. indica var. jaquemontii		T+
A. nilotica ssp. indica var. vediana		T
A. pennatula		T
A. raddiana		T+
A. senegal		T+
A. tortilis		T+

TABLE A-1 (Continued)

Genus/Species	Breeding	Testing
A. tortilis		T+
var. spinocarpa		
Acer (maple)		
A. pseudoplatanus	B	T
A. rubrum	B	
A. saccharinum	B	
A. saccharum	B	
Afrormosia		
A. elata	B	
Agathis (kauri pine)		
A. australis		T
A. dammara		T+
A. lanceolata		T
A. macrophylla		T+
A. moorei		T
A. obtusa		T
A. robusta		T+
A. vitiensis		T
Albizia		
A. caribaea		T
A. falcataria		T
A. procera		T
A. quachepele		T
Alnus (alder)		
A. anisoptera		T
A. cordata		T
A. glutinosa	B	T
A. incana	B	T
A. rubra		T
Anisoptera		
A. spp.		T
Anthocephalus		
A. cadamba		T
Araucaria (monkey puzzle)		
A. angustifolia		T+
A. bidwillii		T
A. cunninghamii	B+	T+
A. hunsteinii	B	T+
Atriplex		
A. repanda		T+
Aucoumea		
A. klaineana	B	T
Azadirachta		
A. indica		T+
Bagassa		
B. guianensis		T

(continued)

TABLE A-1 (*Continued*)

Genus/Species	Breeding	Testing
Betula (birch)		
B. alleghaniensis		T
B. pendula	B	
B. pubescens	B	
B. verrucosa	B	
Bombacopsis		
B. quinata	B	T
Bombax		
B. ceiba	B	T
Calliandra		
C. calothyrsus		T
Callitris (cypress pine)		
C. calcarata		T
C. glauca		T
Calocedrus (northern incense cedar)		
C. decurrens	B	T
Cariniana (jequitiba)		
C. pyriformis	B	
Castanea (chestnut)		
C. dentata	B	T
C. molissima	B	
C. sativa	B	T
Casuarina (she oak)		
C. equisetifolia		T
Cedrela (Chinese cedar)		
C. angustifolia		T
C. mexicana		T +
C. odorata	B	T +
C. tubiflora		T +
Cedrus (cedar)		
C. atlantica		T
C. brevifolia		T
C. deodora		T
C. libani		T
Celtis (hackberry)		
C. occidentalis		T
Chamaecyparis (Alaska cedar)		
C. lawsoniana	B	T
C. obtusa	B	T
Cordia		
C. alliodora	B	T
C. gerascanthus		T
C. goeldiana		T
C. trichotoma		T
Cryptomeria (red cedar)		
C. japonica	B +	T

TABLE A-1 (*Continued*)

Genus/Species	Breeding	Testing
Cunninghamia (Chinese fir)		
C. *lanceolata*	B	T
× *Cupressocyparis*		
C. *leylandii*		T
Cupressus (cypress)		
C. *arizonica*	B	T
C. *benthanii*		T+
C. *japonica*		T
C. *lindleyi*	B	T+
C. *lusitanica*	B+	T+
C. *macrocarpa*	B	T
C. *sempervirens*	B+	T
Dalbergia		
D. *sissoo*	B	T
Didymopanax		
D. *morototoni*		T
Dipterocarpus		
D. spp.		T
Dryobalanops		
D. *aromatica*		T
Endospermum		
D. *macrophyllum*		T
Eucalyptus (eucalyptus)		
E. *alba*	B	T+
E. *apodophylla*		T
E. *biscotata*		T
E. *brassiana*		T+
E. *brassii*		T
E. *bridgesiana*		T
E. *camaldulensis*	B+	T+
E. *citriodora*	B	T+
E. *cloeziana*		T+
E. *cypellocarpa*		T
E. *dalrympleana*		T
E. *deanei*		T
E. *decaisneana*		T
E. *deglupta*	B+	T+
E. *dunnii*	B	T
E. *exerta*		T
E. *fastigata*	B	T
E. *globulus*		T+
E. *gomphocephala*		T
E. *gracilis*		T
E. *grandis*	B+	T+
E. *gunnii*	B	T
E. *houseana*		T

(*continued*)

TABLE A-1 (*Continued*)

Genus/Species	Breeding	Testing
E. *intertexta*		T
E. *kirtoniana*	B	T
E. *macarthurii*		T
E. *maidenii*		T
E. *melliodora*		T
E. *microcorys*		T+
E. *microtheca*		T+
E. *nesophila*		T
E. *nitens*		T
E. *obliqua*		T
E. *paniculata*	B	
E. *pantoleuca*		T
E. *pellita*	B	T+
E. *peltata*		T
E. *pilularis*		T+
E. *pruinosa*		T
E. *ruverentiana*		T
E. *regnans*	B	T+
E. *resinifera*		T
E. *robusta*	B	T
E. *rostrata*		T
E. *rubida*		T
E. *saligna*	B	T+
E. *sieberi*		T
E. *st. johnii*		T
E. *syderoxylon*		T
E. *tereticornis*	B+	T+
E. *torelliana*		T+
E. *umbellata*		T
E. *urophylla*	B	T+
E. *viminalis*	B	T+
Fagus (beech)		
F. *moesiaca*	B	
F. *sylvatica*	B+	T
Fraxinus (ash)		
F. *americana*		T
F. *angustifolia*	B	
F. *excelsior*	B	
F. *floribunda*		T
F. *pennsylvanica*		T
F. *uhdei*		T
Gliciridia		
G. *sepium*		T+
Gmelina		
G. *arborea*	B	T+
Grevillea (silk oak)		
G. *robusta*		T

TABLE A-1   (*Continued*)

Genus/Species	Breeding	Testing
Guazuma		
G. *ulmifolia*		T
Hibiscus		
H. *elatus*	B	
Hura		
H. *crepitans*		T
Ilex (holly)		
I. *paraguariensis*		T
Jacaranda		
J. *copaia*		T
Juglans (walnut)		
J. *nigra*	B	T
J. *regia*	B	T
Juniperus (juniper)		
J. *virginiana*		T
Khaya		
K. *nyasica*	B	
K. *senegalensis*		T
Koelreuteria (golden rain tree)		
K. *paniculata*		T
Larix (larch)		
L. *decidua*	B	T +
L. *kaempferi*	B	T
L. *laricina*	B	T +
L. *leptolepsis*	B +	T
L. *sibirica*		T
Leucaena		
L. *collinsii*		T
L. *diversifolia*	B	T
L. *esculenta*		T
L. *lanceolata*		T
L. *leucocephala*	B +	T +
L. *macrophylla*		T
L. *pulverulenta*	B	T
L. *shannoni*		T
Libocedrus (southern incense cedar)		
L. *decurrens*	B	T
Liquidambar (sweet gum)		
L. *styraciflua*	B	T +
Liriodendron (tulip tree)		
L. *tulipifera*	B	T
Maesopsis		
M. *eminii*	B	
Melia (pride of India)		
M. *azedarach*	B	T +
Metasequoia (Dawn redwood)		
M. *glyptostroboides*		T

(*continued*)

TABLE A-1   (*Continued*)

Genus/Species	Breeding	Testing
Mimosa (mimosa)		
M. scabrella		T
Moringa (horseradish tree)		
M. concanensis	B	
M. oleifera	B	
Morus (mulberry)		
M. alba	B	
M. australis	B	
M. indica	B	
M. laevigata	B	
Nothofagus (southern beech)		
N. alpina		T
N. obliqua		T
N. procera		T
Nyssa (tupelo)		
N. aquatica	B	
N. biflora	B	
Ochroma		
O. lagopus		T
Parkinsonia		
P. acuteata		T
Paulownia (empress tree)		
P. koreana	B	
P. tomentosa	B	
Picea (spruce)		
P. abies	B +	T +
P. engelmannii	B	T
P. glauca	B	T
P. mariana	B	T
P. morinda		T
P. omoric	B	
P. orientalis		T
P. rubens	B	T
P. sitchensis	B +	T +
Pinus (pine)		
P. armandii		T
P. attenuata	B	
P. ayacahuite		T +
P. banksiana	B	T
P. brutia	B	T
P. canariensis		T +
P. caribaea var. *bahamensis*	B +	T +
P. caribaea var. *caribaea*	B +	T +
P. caribaea var. *hondurensis*	B +	T +

TABLE A-1   (Continued)

Genus/Species	Breeding	Testing
P. cembra		T
P. clausa	B	
P. contorta	B+	T+
P. coulteri		T
P. densiflora	B	T
P. douglasiana		T+
P. durangensis		T
P. echinata	B	T+
P. eldarica		T
P. elliottii		T+
var. *densa*		
P. elliottii	B+	T+
var. *elliotii*		
P. engelmannii		T+
P. excelsa		T
P. flexilis		T
P. gerardiana		T
P. greggii		T+
P. halepensis		T
P. halepensis		T
var. *brutia*		
P. hartwegii	B+	T+
P. insularis		T
P. jeffreyi		T
P. kesiya	B+	T+
P. koraiensis	B	
P. lambertiana	B	T
P. lawsonii		T
P. leiophylla		T+
P. luchuensis	B	T
P. lutea		T
P. maestrensis	B	
P. maritima	B	T
P. massoniana		T
P. maximimoi		T
P. merkusii	B+	T+
P. michoacana		T+
P. montezumae		T+
P. monticola	B+	T
P. mugo		T
P. muricata		T+
P. nigra	B+	T
P. nigra	B	
var. *pyramidalis*		
P. nigra	B	T
ssp. *laricio*		
P. oocarpa	B+	T+

(*continued*)

TABLE A-1   (*Continued*)

Genus/Species	Breeding	Testing
P. oocarpa		T
var. *ochoteranai*		
P. palustris	B	T+
P. patula	B+	T+
P. peuce	B	
P. pinaster	B+	T+
P. pinea		T
P. ponderosa		T+
P. pringlei		T
P. pseudostrobus	B+	T+
P. radiata	B+	T+
P. resinosa	B	T
P. rigida	B	T
P. roxburghii	B	T
P. rudis		T+
P. serotina	B	T
P. strobus	B+	T+
P. strobus	B	T
var. *chiapensis*		
P. sylvestris	B+	T+
P. sylvestris		T
var. *mongolica*		
P. tabulaeformis		T
P. taeda	B+	T+
P. taiwanensis		T
P. tecunumanii		T+
P. tenuifolia		T+
P. teocote		T+
P. thunbergii	B	T
P. tropicalis	B	T
P. virginiana	B	T+
P. wallichiana	B	T
Pistacia		
P. kinjux		T
Platanus		
P. occidentalis	B	
P. orientalis	B	T
Populus (poplar)		
P. alba		T
P. augustifolia	B	T
P. ciliata		T
P. deltoides	B	T
P. euphatica	B	
P. fremontii		T
P. glandulosa		T
P. grandidentata		T
P. koreana		T

TABLE A-1 (*Continued*)

Genus/Species	Breeding	Testing
P. laurifolia		T
P. maximowiczii		T
P. nigra		T
P. simonii		T
P. szechuanica		T
P. tracamahaca		T
P. tremula		T
P. tremuloides	B	T
P. trichocarpa	B	
P. wilsonii		T
P. yunnanensis		T
Prosopis		
P. chilensis		T+
P. cineraria		T
P. juliflora		T
P. spicigera		T
P. tamarugo		T+
Prunus		
P. avium	B	T
P. serotina	B	
Pseudotsuga (Douglas fir)		
P. menziesii	B+	T+
Pterocarpus		
P. santalinus	B	
Quercus (oak)		
Q. aegilops		T
Q. alba		T
Q. borealis	B	
Q. coccinea		T
Q. falcata	B	
var. *pagodaefolia*		
Q. ilex	B+	
Q. infectoria		T
Q. palustris	B	
Q. petraea	B	
Q. robur	B	T
Q. rubra		T+
Robinia		
R. pseudoacacia	B	T
Salix (willow)		
S. alba	B	
S. argentinensis	B	
S. babylonica		
var. *sacramenta*	B	
S. humboldiana	B	
Salmalia		
S. malabaricum (*Bombax*)		T

(*continued*)

TABLE A-1 (*Continued*)

Genus/Species	Breeding	Testing
Schizolobium		
S. parahybum		T
Sclerolobium		
S. paniculatum		T
Sequoia (redwood)		
S. sempervirens	B	T
Sequoiadendron (giant sequoia)		
S. giganteum	B	T
Sesbania		
S. grandiflora		T
Shorea		
S. spp.		T
Simaruba		
S. amara (*Quassia amara*)		T
Sorbus (whitebeam; mountain ash)		
S. ancuparia		T
S. domestica		T
S. torminalis		T
Spathodea (African tulip tree)		
S. campanulata		T
Swietenia		
S. humilis		T
S. macrophylla		T+
S. mahogani		T
Tabebuia		
T. azellanedae		T
T. rosea	B	
Taxodium (swamp cypress)		
T. mucronatum		T
Taxus (yew)		
T. brevifolia		T
Tectona (teak)		
T. grandis	B+	T+
Terminalia		
T. brassii		T
T. ivorensis	B	
T. superba	B	T+
Thuja		
T. plicata	B	T
Tilia (lime, linden, basswood)		
T. argentea	B	
T. cordata	B	
T. platyphyllos	B	
Toona		
T. ciliata	B	T
Triplochiton (obeche)		
T. scleroxylon	B	

TABLE A-1 *(Continued)*

Genus/Species	Breeding	Testing
Tsuga (hemlock)		
T. heterophylla		T
T. mertensiana		T
Ulmus (elm)		
U. americana	B	
U. campestris	B	
U. pumila		T

B

Literature Survey of Threatened Provenances or Species

The categories in Table B-1 that denote the degree of threat to the provenances and species are derived from the classification system used by the International Union for the Conservation of Nature and Natural Resources (IUCN, 1978a). The categories and their total numbers from the list are the following:

• Ex (extinct): not found after repeated searches of known and likely areas, 6.

• E (endangered): in danger of extinction and with survival being unlikely if the causal factors continue to operate, 147.

• V (vulnerable): believed likely to move into endangered category in the near future if the causal factors continue to operate, 156.

• R (rare): small populations that are not currently endangered or vulnerable but that are at risk, 42.

• T (threatened or indeterminate): threatened provenances or species about which there is insufficient information to say which of the four categories above is appropriate. It also includes taxa with threatened provenances, 173.

An asterisk (*), of which there are 209, indicates the species is not known to be included in any conservation program. In some cases, more than one category applies to a provenance or species.

Many of the species on this list are common, but they have provenances identified as being threatened to some degree. Further, the information is based on a literature survey and does not represent an exhaustive list of all pertinent tree species and provenances.

In addition to the IUCN, acronyms noted in column three are FAO, Food and Agriculture Organization, and SARH–INIF, Secretaría de Agricultura y Recursos Hidraulicos–Instituto Nacional de Investigación Forestal (Secretariat of Agriculture and Hydraulic Resources–National Institute of Forest Research).

TABLE B-1   Species or Provenances Reported in the Literature as Threatened in Whole or in Some Significant Part

Genus/Species	Status of Provenances or Species	Source(s)
Abies		
A. concolor	E	SARH-INIF, 1979
A. guatemalensis	E	FAO, 1986a
A. nebrodensis	E	FAO, 1986a
A. numidica	V	FAO, 1986a
Acacia		
A. albida	V	FAO, 1986a,b
A. aphylla	E*	IUCN, 1978b
A. caven	V	FAO, 1986a
A. gassei	E	FAO, 1986b
A. nilotica ssp. *adstringens*	V	FAO, 1986b
A. nilotica ssp. *tomentosa*	V	FAO, 1986b
A. peuce	V	IUCN, 1978b
A. senegal	V	FAO, 1986b
A. tortilis ssp. *raddiana*	V	FAO, 1986a,b
A. tortilis ssp. *syncocarpa*	T	FAO, 1985
A. tortilis ssp. *tortilis*	V	FAO, 1986a,b
Achras		
A. zapota (*Manilkara zapota*)	T	FAO, 1985
Achyranthes		
A. mangarevica	Ex/E*	IUCN, 1978b
Adansonia		
A. digitata	V	FAO, 1986b
Afraegle		
A. asoo	V*	FAO, 1984a
Afrosersalisia		
A. afzelii	V*	FAO, 1984a
Afzelia		
A. africana	V	Palmberg, 1987
A. bipendensis	V	Palmberg, 1987
A. pachyloba	V	Palmberg, 1987

TABLE B-1 *(Continued)*

Genus/Species	Status of Provenances or Species	Source(s)
Agathis		
A. borneensis	V	FAO, 1984a
Ailanthus		
A. fordii	R*	IUCN, 1978b
Albizia		
A. arunachalensis	T*	FAO, 1981
A. caribaea	T	FAO, 1985
A. gamblei	T*	FAO, 1981
Alluaudia		
A. procera	E	FAO, 1977
Alnus		
A. acuminata	V	FAO, 1986a; Palmberg, 1987
A. jorullensis	V	FAO, 1984a, 1986a
Amburana		
A. cearensis	V	FAO, 1984a
Anadenanthera		
A. macrocarpa	V	FAO, 1986a
Andira		
A. inermis	E*	FAO, 1984a
Aniba		
A. duckei	E*	FAO, 1986a
A. spp.	V	FAO, 1984a
Anogeissus		
A. leiocarpa	V	FAO, 1986b
Aquilaria		
A. malaccensis	R	FAO, 1984a
Araliopsis		
A. soyauxii	V*	FAO, 1984a
Araucaria		
A. angustifolia	V	FAO, 1986a
A. araucana	T*	FAO, 1981
A. cunninghamii	V	FAO, 1986a
A. hunsteinii	V	FAO, 1986a
Arbutus		
A. canariensis	V*	IUCN, 1978b
Aspidosperma		
A. olivaceum	T	FAO, 1985
A. polyneuron	E	FAO, 1986a
A. pyrifolium	T	FAO, 1985
Astronium		
A. fraxinifolium	T	FAO, 1985
A. gracile	T	FAO, 1985
A. urundeuva	E	FAO, 1986a
Atriplex		
A. repanda	V	FAO, 1981, 1986a

(continued)

TABLE B-1 (Continued)

Genus/Species	Status of Provenances or Species	Source(s)
Aucoumea		
A. klaineana	E*	FAO, 1977, 1984a
Badula		
B. crassa	E*	IUCN, 1978b
Baikiaea		
B. plurijuga	T	Palmberg, 1987
Balanites		
B. aegyptiaca	V	FAO, 1986b
Balfourodendron		
B. riedelianum	V	FAO, 1986a
Basiloxylon		
B. excelsum	T*	FAO, 1981
Batocarpus		
B. costaricensis	T*	FAO, 1981
Beilschmiedia		
B. pseudomicropora	T*	FAO, 1981, 1986a
Berlinia		
B. confusa	E	Palmberg, 1987
Bertholletia		
B. excelsa	V	FAO, 1981, 1986a
Betula		
B. uber	E*	IUCN, 1978b
B. utilis	T*	FAO, 1981
Blighia		
B. sapida	V*	FAO, 1984a
B. unijugata	V*	FAO, 1984a
Bombacopsis		
B. quinata	V	FAO, 1986a
Borassus		
B. aethipium	V	FAO, 1986b
B. flabellifer	V	FAO, 1986b
Boswellia		
B. freinana	E	FAO, 1986b
B. sacra	E	FAO, 1986b
Brachylaena		
B. hutchinsii	E	FAO, 1977, 1986a
Brachystegia		
B. eurycoma	V*	FAO, 1984a
B. kennedyi	V*	FAO, 1984a
Buddleia		
B. spp.	V	FAO, 1984a
Bursera		
B. aloexylon	R/E*	FAO, 1981
B. leptophleos	T	FAO, 1985

TABLE B-1 *(Continued)*

Genus/Species	Status of Provenances or Species	Source(s)
Burttdavya		
B. nyasica	E*	FAO, 1977
Caesalpinia		
C. dalei	E*	FAO, 1984a
C. leiostachia	T	FAO, 1985
Calamus		
C. caesius	E	FAO, 1984a
C. manan	E	FAO, 1984a; Palmberg, 1987
Calophyllum		
C. brasiliense	V	FAO, 1984a
Calpocalyx		
C. heitzii	E	FAO, 1984a; Palmberg, 1987
Camellia		
C. crapnelliana	E*	IUCN, 1978b
C. granthamiana	T*	IUCN, 1978b
Carapa		
C. grandiflora	V*	FAO, 1984a
Cariniana		
C. estrelensis	T	FAO, 1985
C. legulis	T	FAO, 1985
Carpinus		
C. caroliniana	R/E*	FAO, 1981
Caryocar		
C. costaricense	T*	FAO, 1981
C. vellosum	T	FAO, 1985
Casuarina		
C. fibrosa	E*	IUCN, 1978b
Cecropia		
C. cinerea	T	Palmberg, 1987
C. peltata	V	FAO, 1984a
Cedrelinga		
C. catenaeformis	V	FAO, 1984a; Palmberg, 1987
Cedrela		
C. fissilis	V	FAO, 1986a
C. odorata	V	FAO, 1981, 1986a; Palmberg, 1987
Cedrus		
C. libani	V	FAO, 1986a
Ceiba		
C. pentandra	R/E*	FAO, 1981

(continued)

TABLE B-1 (Continued)

Genus/Species	Status of Provenances or Species	Source(s)
Celtis		
C. aetnensis	T*	FAO, 1981
Centrolobium		
C. robustum	T	FAO, 1985
Ceratonia		
C. sp. nov.	E*	IUCN, 1978b
Cercidiphyllum		
C. japonicum	V	FAO, 1986a
Cercis		
C. canadensis	R*/E	FAO, 1981
Chlorophora		
C. excelsa	E	FAO, 1986a
Chorisia		
C. speciosa	V	FAO, 1984a; Palmberg, 1987
Chyrantodendron		
C. pentadactylon	R/E*	FAO, 1981
Cladrastis		
C. lutea	V*	IUCN, 1978b
Colubrina		
C. glandulosa	T	FAO, 1985
Copaifera		
C. langesdorfii	T	FAO, 1985
C. officinalis	V	FAO, 1984a
C. religiosa	E	FAO, 1984a; Palmberg, 1987
Cordeauxia		
C. edulis	E	FAO, 1986b
Cordia		
C. alliodora	R/E*	FAO, 1981
C. eleagnoides	R/E*	FAO, 1981
C. millenii	V*	FAO, 1981
Cordyline		
C. kaspar	R*	IUCN, 1978b
Cornus		
C. disciflora	R/E*	FAO, 1981
Crudia		
C. klainei	E	Palmberg, 1987
Cryptosepalum		
C. staudtii	V*	FAO, 1984a
Cupressus		
C. atlantica	E	FAO, 1986a
C. benthamii	R/E*	FAO, 1981
C. dupreziana	E	FAO, 1986a; IUCN, 1978b

TABLE B-1   (*Continued*)

Genus/Species	Status of Provenances or Species	Source(s)
C. forbesii	R/E*	FAO, 1981
C. guadalupensis	R/E*	FAO, 1981
C. lusitanica	E	FAO, 1985
C. macrocarpa	E*	IUCN, 1978b
C. montana	E	SARH-INIF, 1979
C. sempervirens	T*	FAO, 1981; IUCN, 1978b
Cynometra		
C. hemitomophylla	T*	FAO, 1981
Dacryodes		
D. buettneri	E	FAO, 1984a; Palmberg, 1987
D. igaganga	E	FAO, 1984a; Palmberg, 1987
D. klaineana	E	Palmberg, 1987
Dalbergia		
D. nigra	E	FAO, 1986a
D. sisso	V	Palmberg, 1987
Daniellia		
D. africana	V*	FAO, 1984a
Dendrosicyos		
D. socotranus	V*	IUCN, 1978b
Dialium		
D. bipindense	V*	FAO, 1984a
Didelotia		
D. africana	V*	FAO, 1984a
D. unifoliata	V*	FAO, 1984a
Didymopanax		
D. morototoni	V	FAO, 1986a
Diospyros		
D. cacharensis	T*	FAO, 1981
D. hemiteles	E	FAO, 1986a; IUCN, 1978b
Dipterocarpus		
D. spp.	T	FAO, 1984a
Dipteryx		
D. alata	V	FAO, 1986a
Dipthysa		
D. robinoides	T*	FAO, 1981
Dirachma		
D. socotrana	E*	IUCN, 1978b
Donella (Chrysophyllum)		
D. pruniformis	V*	FAO, 1984a
Dorstenia		
D. gigas	T*	IUCN, 1978b

(*continued*)

TABLE B-1 (Continued)

Genus/Species	Status of Provenances or Species	Source(s)
Dracaena		
D. draco	V*	IUCN, 1978b
D. ombet	T*	IUCN, 1978b
Dryobalanops		
D. aromatica	V	Palmberg, 1987
Drypetes		
D. caustica	E*	IUCN, 1978b
Durio		
D. spp.	T	FAO, 1984a
Dyera		
D. costulata	V	FAO, 1984a; Palmberg, 1987
Dysoxylum		
D. reticulatum	T*	FAO, 1981
Elliotia		
E. racemosa	E*	IUCN, 1978b
Engelhardia		
E. pterocarpa	T*	FAO, 1981
Entandrophragma		
E. angolense	V*	FAO, 1986a
E. utile	V	Palmberg, 1987
Erythrina		
E. mildbraedii	V*	FAO, 1984a
Esenbeckia		
E. leiocarpa	E	FAO, 1986a; Palmberg, 1987
Eucalyptus		
E. acies	T*	FAO, 1981
E. alpina	T*	FAO, 1981
E. aproximans ssp. aproximans	T*	FAO, 1981
E. aquilina	T*	FAO, 1981
E. archeri	T	FAO, 1981
E. argophloia	E	FAO, 1981; IUCN, 1978b
E. badjensis	T	FAO, 1981
E. baeuerlenii	T	FAO, 1981
E. bakeri	T	FAO, 1981
E. barberi	T	FAO, 1981
E. beardiana	T	FAO, 1981
E. benthamii	T	FAO, 1981
E. brockwayi	T*	FAO, 1981
E. burdettiana	T*	FAO, 1981
E. burgessiana	T*	FAO, 1981
E. caesia	T*	FAO, 1981

TABLE B-1 (*Continued*)

Genus/Species	Status of Provenances or Species	Source(s)
E. calcicola	T*	FAO, 1981
E. camfieldii	T*	FAO, 1981
E. carnabyi	E*	FAO, 1981
E. cneorifolia	T*	FAO, 1981
E. conglomerata	T*	FAO, 1981
E. cordata	T*	FAO, 1981
E. coronata	T*	FAO, 1981
E. crenulata	E*	FAO, 1981; IUCN, 1978b
E. cupularis	T*	FAO, 1981
E. curtisii	V	FAO, 1981; IUCN, 1978b
E. deglupta	V	FAO, 1986a
E. dendromorpha	T	FAO, 1981
E. desmondensis	T*	FAO, 1981
E. ficifolia	T	FAO, 1981
E. fitzgeraldii	T*	FAO, 1981
E. froggattii	E*	FAO, 1981; IUCN, 1978b
E. georgei	T*	FAO, 1981
E. globulus ssp. globulus	V	FAO, 1986a
E. gregsoniana	T	FAO, 1981
E. halophila	T	FAO, 1981
E. imlayensis	T*	FAO, 1981
E. insularis	T	FAO, 1981
E. johnsoniana	T*	FAO, 1981
E. kartzoffiana	T*	FAO, 1981
E. kitsoniana	T	FAO, 1981
E. kruseana	T	FAO, 1981
E. lane-poolei	T	FAO, 1981
E. lansdowneana ssp. lansdowneana	T	FAO, 1981
E. largeana	T*	FAO, 1981
E. luehmanniana	T*	FAO, 1981
E. megacornuta	T	FAO, 1981
E. michaeliana	T*	FAO, 1981
E. mitchelliana	T	FAO, 1981
E. morrisbyi	T	FAO, 1981
E. neglecta	T	FAO, 1981
E. nigra	T	FAO, 1981
E. olsenii	T	FAO, 1981
E. paliformis	T	FAO, 1981
E. parvifolia	T	FAO, 1981
E. pendens	T*	FAO, 1981

(*continued*)

TABLE B-1 (*Continued*)

Genus/Species	Status of Provenances or Species	Source(s)
E. *pulverulenta*	T	FAO, 1981
E. *pumila*	T	FAO, 1981
E. *remota*	E	FAO, 1981
E. *rhodantha*	E*	FAO, 1981
E. *risdonii*	T	FAO, 1981
E. *rummeryi*	T	FAO, 1981
E. *saxatilis*	T*	FAO, 1981
E. *scoparia*	T	FAO, 1981
E. *sepulcralis*	T	FAO, 1981
E. *squamosa*	T*	FAO, 1981
E. *steedmanii*	Ex*	FAO, 1981; IUCN, 1978b
E. *stenostoma*	T*	FAO, 1981
E. *stogtei*	T*	FAO, 1981
E. *stowardii*	T*	FAO, 1981
E. *sturgissiana*	T*	FAO, 1981
E. *tetrapleura*	T*	FAO, 1981
E. *triflora*	T	FAO, 1981
E. *woodwardii*	T	FAO, 1981
E. *xanthonema*	T	FAO, 1981
E. *yarraensis*	T	FAO, 1981
Euphorbia		
E. *wakefieldii*	E*	IUCN, 1978b
Fagara (*Zanthoxylum*)		
F. *xamthoxyloides*	V*	FAO, 1984a
Fagus		
F. *longipetiolata*	V	FAO, 1986a
F. *mexicana*	R/E*	FAO, 1981
Ficus		
F. spp.	V	FAO, 1984a
Fitzroya		
F. *cupressoides*	T*	FAO, 1981
Franklinia		
F. *alatamaha*	Ex*	IUCN, 1978b
Freziera		
F. *forerorum*	E*	IUCN, 1978b
Gambeya (*Chrysophyllum*)		
G. *africana*	V	Palmberg, 1987
G. *beguei*	V	Palmberg, 1987
G. *boukokoensis*	V	Palmberg, 1987
G. *lacourtiana*	V	Palmberg, 1987
G. *perpulchra*	V	Palmberg, 1987
Ganophyllum		
G. *giganteum*	V*	FAO, 1984a
Garcinia		
G. *kola*	V*	FAO, 1984a

TABLE B-1   (*Continued*)

Genus/Species	Status of Provenances or Species	Source(s)
Gaultheria		
G. seshagiriana	T*	FAO, 1981
Genipa		
G. americana	E	FAO, 1985
Gigantochloa		
G. scortechinii	T*	FAO, 1984a
Gigasiphon		
G. macrosiphon	E	FAO, 1986a; IUCN, 1978b
Gilletiodendron		
G. mildbraedii	V*	FAO, 1984a
G. pierreanum	E	Palmberg, 1987
G. preussii	V*	FAO, 1984a
Gluema		
G. ivorensis	E	Palmberg, 1987
Glyptostrobus		
G. lineatus	Ex/E	Palmberg, 1987
G. pensilis	E	FAO, 1985
Gonystylus		
G. bancanus	V*	FAO, 1984a
Gossweilerodendron		
G. balsamiferum	V	FAO, 1986a
Guaiacum		
G. sanctum	T*	FAO, 1981
Guarea		
G. longipetiola	T*	FAO, 1981
Guibourtia		
G. ehie	E	Palmberg, 1987
Gymnocladus		
G. assamicus	T*	FAO, 1981
Gymnostemon		
G. zaizou	R	FAO, 1986a
Heritiera		
H. longipetiolata	E*	IUCN, 1978b
Hevea		
H. brasiliensis	V	FAO, 1984a
Hibiscadelphus		
H. giffardianus	E*	IUCN, 1978b
H. wilderianus	Ex*	IUCN, 1978b
Homalium		
H. longistylum	V*	FAO, 1984a
Hymenaea		
H. courbaril	V	Palmberg, 1987
H. stilbocarpa	T	FAO, 1985
Hyphaene		
H. thebaica	V	FAO, 1986b

(*continued*)

TABLE B-1   (Continued)

Genus/Species	Status of Provenances or Species	Source(s)
Ilex		
I. paraguariensis	V	FAO, 1986a
Indiospermum		
I. australiense	V*	IUCN, 1978b
Inidosculus		
I. phyllocanthus	T	FAO, 1985
Inocarpus		
I. edulis	T*	FAO, 1981
Irvingia		
I. gabonensis	V	FAO, 1986a
Intsia		
I. palembanica	V	FAO, 1984a
Joannesia		
J. princeps	E	FAO, 1986a
Juglans		
J. neotropica	V	FAO, 1984a
Juniperus		
J. bermudiana	V	FAO, 1986a; IUCN, 1978b
J. californica	R/E*	FAO, 1981
J. comitana	E	FAO, 1985
J. deppeana var. *pachyphlaea*	R/E*	FAO, 1981
J. gamboana	E	FAO, 1985
J. macropoda	T*	FAO, 1981
J. phoenicea	T*	FAO, 1981
J. procera	E	FAO, 1986a
J. standleyi	R/E*	FAO, 1981
Jura		
J. crepitans	V	FAO, 1984a
Kantou		
K. guereensis	E	Palmberg, 1987
Khaya		
K. senegalensis	V	FAO, 1986a; Palmberg, 1987
Koompassia		
K. excelsa	T	FAO, 1984a
K. malaccensis	T	FAO, 1984a
Lebronnecia		
L. kokioides	R	IUCN, 1978b
Lecythis		
L. ampla	T*	FAO, 1981
L. pisonis	E	FAO, 1985
Leplaea		
L. mayombensis	V*	FAO, 1984a

TABLE B-1 (*Continued*)

Genus/Species	Status of Provenances or Species	Source(s)
Letestua		
L. *durissima*	E	Palmberg, 1987
Libocedrus		
L. *decurrens*	R/E	FAO, 1981; SARH-INIF, 1979
Librevillea		
L. *klainei*	V*	FAO, 1984a
Liquidambar		
L. *styraciflua*	V	FAO, 1986a
Lithocarpus		
L. *kamengensis*	T*	FAO, 1981
Loesenera		
L. *talbotii*	V*	FAO, 1984a
Lophira		
L. *alata*	V	Palmberg, 1987
Lovoa		
L. *swynnertonii*	R	FAO, 1986a
Loxopterygium		
L. *huasango*	V	FAO, 1984a
Luculia		
L. *grandifolia*	T*	FAO, 1981
Machaerium		
M. *villosum*	V	FAO, 1986a
Magnistipula		
M. *butayei*	V*	FAO, 1984a
Majidea		
M. *foresti*	V*	FAO, 1984a
Manilkara		
M. *fouilloyana*	V*	FAO, 1984a
Medusagyne		
M. *oppositifolia*	E*	IUCN, 1978b
Melanoxylum		
M. *brauna*	E	FAO, 1985
Miconia		
M. *cinnamomifolia*	T	FAO, 1985
Millettia		
M. *laurentii*	V*	FAO, 1984a
Mimosa		
M. *caesalpiniaefolia*	V	FAO, 1986a
M. *verrucosa*	V	FAO, 1986a
Mimusops		
M. *angel*	E	FAO, 1986b
M. *dagan*	E	FAO, 1986b
Mitrephora		
M. *harae*	T*	FAO, 1981

(*continued*)

TABLE B-1   (Continued)

Genus/Species	Status of Provenances or Species	Source(s)
Monopetalanthus		
M. hedinii	V*	FAO, 1984a
Mycrocarpus		
M. frondosus	T	FAO, 1985
Myroxylon		
M. balsamum	T*	FAO, 1981
Neobalanocarpus		
N. heimii	V	FAO, 1984a; Palmberg, 1987
Neowawraea		
N. phyllanthoides	E*	IUCN, 1978b
Nesogordonia		
N. papaverifera	V	FAO, 1986a
Nuxia		
N. congesta	V*	FAO, 1984a
Ochroma		
O. ochroma lagopus	T	FAO, 1985
Ochthocosmus		
O. calotthyrsus	V*	FAO, 1984a
Ocotea		
O. porosa	E	FAO, 1986a
O. pretiosa	T	FAO, 1985
Oldfieldia		
O. africana	V*	FAO, 1984a
Olea		
O. laperrinei	V*	IUCN, 1978b
Olneya		
O. tesota	R/E*	FAO, 1981
Ostrya		
O. virginiana	R/E*	FAO, 1981
Palaquium		
P. gutta	V	FAO, 1984a
Paratecoma		
P. peroba	T	FAO, 1985
Parkia		
P. biglobosa	T	FAO, 1984b
P. speciosa	V	FAO, 1984a; Palmberg, 1987
Pausinystalia		
P. johimbe	V*	FAO, 1984a
Pentadesma		
P. grandifolia	V*	FAO, 1984a
Pericopsis		
P. elata	V	FAO, 1981, 1986a

TABLE B-1 (Continued)

Genus/Species	Status of Provenances or Species	Source(s)
Persea		
P. theobromifolia	E*	IUCN, 1978b
Picea		
P. chihuahuana	E	FAO, 1977, 1981
P. mexicana	E	FAO, 1977, 1981
Pilgerodendron		
P. uviferum	T*	FAO, 1981
Pinus		
P. arizonica	T	FAO, 1985
P. armandii	E	FAO, 1986a
var. *amamiana*		
P. attenuata	R/E	FAO, 1985; Theisen, 1980
P. ayacahuite	E	FAO, 1985
P. bhutanica	T*	FAO, 1981
P. cembroides	E	FAO, 1985
var. *edulis*		
P. contorta	R/E	FAO, 1981
var. *latifolia*		
P. coulteri	R/E	FAO, 1981; SARH-INIF, 1979
P. culminicola	R/E*	FAO, 1981
P. douglasiana	T	FAO, 1985
P. eldarica	E	FAO, 1986a
P. greggii	T	FAO, 1985
P. halepensis	T	FAO, 1981
P. jeffreyi	R/E	FAO, 1981; SARH-INIF, 1979
P. koraiensis	V	FAO, 1986a
P. maximartinezii	E	FAO, 1977, 1981
P. michoacana	T	FAO, 1985
P. monophylla	R/E	FAO, 1981
P. montezumae	T	FAO, 1985
P. morrisonicola	E	FAO, 1977
P. muricata	R/E	FAO, 1981
P. oaxacana	T	FAO, 1985
P. oocarpa	T	FAO, 1985
P. patula	V	FAO, 1986a
subsp. *tecunumanii*		
(*P. tecunumanii*)		
P. pentaphylla	V	FAO, 1986a
P. pseudostrobus	V	FAO, 1986a
P. pseudostrobus/ tenuifolia	T	FAO, 1985

(continued)

TABLE B-1 (*Continued*)

Genus/Species	Status of Provenances or Species	Source(s)
P. pseudostrobus var. *oaxacana*	E	FAO, 1986a
P. radiata	E	FAO, 1986a
P. radiata var. *binata*	E	FAO, 1981; SARH-INIF, 1979
P. rzedowski	R*	FAO, 1981
P. strobus var. *chiapensis* (*P. chiapensis*)	R/E	FAO, 1977, 1981
P. tenuifolia	E	FAO, 1977
Piptadenia		
P. macrocarpa	T	FAO, 1985
P. peregrina	V	FAO, 1986a
Pistacia		
P. atlantica	T*	FAO, 1981
Pithecellobium		
P. parvifolium	T	FAO, 1985
Pittosporum		
P. dallii	E*	IUCN, 1978b
Platanus		
P. orientalis	V	FAO, 1986a
P. wrightii	T*	FAO, 1981
Plathymenia		
P. foliosa	E	FAO, 1986a
Platonia		
P. insignis	T	FAO, 1985
Podocarpus		
P. costalis	T*	FAO, 1981
P. lambertii	T	FAO, 1985
P. matudae	R/E*	FAO, 1981
P. milanjianus	V*	FAO, 1984a
P. parlatorei	T*	FAO, 1981
P. reikei	R/E*	FAO, 1981
P. rospigliosii	V	FAO, 1984a
P. utilior	V	FAO, 1984a
Polylepis or *Polylepsis*		
P. racemosa	V	FAO, 1984a; Palmberg, 1987
Populus		
P. gamblei	T*	FAO, 1981
P. ilicifolia	E	FAO, 1986a
Prerocarpus		
P. lucens	V	FAO, 1986b
Prosopis		
P. africana	V	FAO, 1986b

TABLE B-1 (*Continued*)

Genus/Species	Status of Provenances or Species	Source(s)
P. cineraria	V	FAO, 1981, 1986a; Palmberg, 1987
P. juliflora	V	FAO, 1981, 1984a
P. pubescens	T	FAO, 1985
Prunus		
P. africana	V*	FAO, 1984a
P. gravesii	E*	IUCN, 1978b
Pseudotsuga		
P. flahaulti	E	FAO, 1985
P. gausseni	E	FAO, 1986a
P. menziesii	T	FAO, 1981
P. sinensis	R	FAO, 1986a
Pterogyne		
P. nitens	V	FAO, 1986a
Pterygota		
P. mildbraedii	V*	FAO, 1984a
Punica		
P. protopunica	E*	IUCN, 1978b
Pycnanthus		
P. morcalianus	V*	FAO, 1984a
Quercus		
Q. copeyensis	T*	FAO, 1981
Radermachera		
R. sinica	T*	FAO, 1981
Raphia		
R. sudanica	E	FAO, 1986b
Rhizophora		
R. spp.	T	FAO, 1984a
Rhododendron		
R. dalhousiae	T*	FAO, 1981
R. tawangensis	T*	FAO, 1981
Rothmannia		
R. lujae	V*	FAO, 1984a
Salix		
S. bhutanensis	T*	FAO, 1981
S. goodingii	T*	FAO, 1981
S. silicola or silicicola	R*	IUCN, 1978b
Santalum		
S. fernandezianum	Ex*	IUCN, 1978b
Schinopsis		
S. brasiliensis	V	FAO, 1986a
Scytopetalum		
S. klaineanum	V*	FAO, 1986a

(*continued*)

TABLE B-1   (Continued)

Genus/Species	Status of Provenances or Species	Source(s)
Serianthes		
S. nelsonii	E*	IUCN, 1978b
Shorea		
S. curtisii	T	FAO, 1984a
S. gratissima	V	Palmberg, 1987
S. platyclados	T	FAO, 1984a
Sindoropsis		
S. letestui	V	FAO, 1984a
Sophora		
S. fernandeziana	V*	IUCN, 1978b
S. masafuerana	E*	IUCN, 1978b
S. toromiro	E*	IUCN, 1978b
Spondias		
S. macrocarpa	E	FAO, 1985
S. purpura	E	FAO, 1985
S. tuberosa	T	FAO, 1985
Stipa		
S. tenacissima	T*	FAO, 1981
Stuhlmannia		
S. moavi	E	FAO, 1986a
Swartzia		
S. cubensis	E	FAO, 1985
Sweetia		
S. panamensis	E	FAO, 1985
Swietenia		
S. humilis	T*	FAO, 1981
S. macrophylla	V	FAO, 1984a; Palmberg, 1987
Syagrus		
S. coronata	T	FAO, 1985
Symaplacos		
S. glauca	T*	FAO, 1981
Symplocos		
S. coynadense	T*	FAO, 1981
S. latiflora	T*	FAO, 1981
S. oligandra	T*	FAO, 1981
S. sessilis	T*	FAO, 1981
Syzygium		
S. assamicum	T*	FAO, 1981
Tabebuia		
T. avellanedae	E	FAO, 1985
T. cassinoides	E	FAO, 1985
T. impetiginosa	V	FAO, 1986a
T. rosea	E	FAO, 1985
T. serratifolia	V	FAO, 1984a

TABLE B-1 (*Continued*)

Genus/Species	Status of Provenances or Species	Source(s)
Tachigali		
T. *versicolor*	T*	FAO, 1981
Taiwania		
T. *cryptomerioides*	V	FAO, 1986a
T. *flousiana*	E	FAO, 1986a
Taxus		
T. *globosa*	R/E*	FAO, 1981
Tectona		
T. *hamiltoniana*	R*	FAO, 1986a
T. *philippinensis*	E*	FAO, 1986a
Terminalia		
T. *amazonica*	E	FAO, 1985
T. *ivorensis*	R	
Testulea		
T. *brevipaniculata*	E	FAO, 1984a
T. *gabonensis*	T	FAO, 1984a; Palmberg, 1987
Testules		
T. *gabonensis*	V*	FAO, 1984a
Tetraberlinia		
T. *polyphylla*	V*	FAO, 1984a
Tetrataxis		
T. *salicifolia*	E*	IUCN, 1978b
Tieghemella		
T. *africana*	R	Palmberg, 1987
Tilia		
T. *mexicana*	R/E*	FAO, 1981
Toubaouate (*Didelotia*)		
T. *brevipaniculata*	E	Palmberg, 1987
Trema		
T. *micrantha*	V	Palmberg, 1987
Triplochiton		
T. *scleroxylon*	V	Palmberg, 1987
Trochetia		
T. *erythroxylon*	E*	FAO, 1981; IUCN, 1978b
Trochetiopsis		
T. spp.	T*	FAO, 1981
Ulmus		
U. *wallichiana*	E	FAO, 1986a; IUCN, 1978b
Vantanea		
V. *barbourii*	T*	FAO, 1981
Vateria		
V. *seychellarum*	E*	IUCN, 1978b

(*continued*)

TABLE B-1 (*Continued*)

Genus/Species	Status of Provenances or Species	Source(s)
Vepris		
V. glandulosa (syn. *Tecleopsis glandulosa*)	E	FAO, 1986a; IUCN, 1978b
Virola		
V. spp.	V	FAO, 1984a
Vitellaria or *Vitellaila*		
V. paradoxa	V	FAO, 1986b
Zanthoxylum		
Z. paniculatum	E*	IUCN, 1978b
Zeyheria		
Z. tuberculosa	V	FAO, 1986a; Palmberg, 1987
Ziziphus		
Z. joazeiro	T	FAO, 1985

SOURCES:

Food and Agriculture Organization (FAO). 1977. Report of the Fourth Session of the FAO Panel of Experts on Forest Gene Resources. FO:FGR/4/Rep. Rome, Italy: Food and Agriculture Organization.

FAO. 1981. Report of the Fifth Session of the FAO Panel of Experts on Forest Gene Resources. Rome, Italy: Food and Agriculture Organization.

FAO. 1984a. A Guide to In-situ Conservation of Genetic Resources for Tropical Woody Species. FORGEN/MISC/84/2. Rome, Italy: Food and Agriculture Organization.

FAO. 1984b. In Situ Conservation of Wild Plant Genetic Resources: A Status Review and Action Plan (Draft) Background Document for the First Session of the FAO Commission on Plant Genetic Resources. FORGEN/MISC/84/3. Rome, Italy: Food and Agriculture Organization.

FAO. 1985. Report of the Sixth Session of the FAO Panel of Experts on Forest Gene Resources. FO:FGR/5/Rep. Rome, Italy: Food and Agriculture Organization.

FAO. 1986a. Databook on Endangered Tree and Shrub Species and Provenances. Forestry Paper 77. Rome, Italy: Food and Agriculture Organization.

FAO. 1986b. Selection and genetic improvement of indigenous and exotic multipurpose woody species; including seed collection, handling, storage, and exchange. Unpublished paper presented at the Research Planning Workshop for Africa, Sahelian and North Sudanian Zones, International Union of Forestry Research Organizations and the Forestry Department of Kenya, Nairobi, January 9–15, 1986.

TABLE B-1 *(Continued)*

International Union for the Conservation of Nature and Natural Resources (IUCN). 1978a. Categories, Objectives, and Criteria for the Protected Areas. Gland, Switzerland: International Union for the Conservation of Nature and Natural Resources.

IUCN. 1978b. The IUCN Plant Red Data Book. Gland, Switzerland: International Union for the Conservation of Nature and Natural Resources.

Mabberley, D. J. 1989. The Plant-Book. New York: Cambridge University Press. (Consulted to address several taxonomic discrepancies in the source literature.)

Palmberg, C. 1987. Conservation of genetic resources of woody species. Unpublished paper presented at the Symposium on Silvaculture and Genetic Improvement, Centro Investigación y Estudios Forestales, Buenos Aires, April 6–10, 1987.

Secretaría de Agricultura y Recursos Hidraulicos–Instituto Nacional de Investigación Forestal. 1979. Memoria: III Simposio Binacional Sobre el Medio Ambiente del Golfo de California-Ecodesarrollo. Publicación Especial No. 14. Mexico: Secretaría de Agricultura y Recursos Hidraulicos.

Theisen, P. A. 1980. Maintenance of Genetic Diversity and Gene Pool Conservation of Commercial Tree Species in the Pacific Northwest Region. Washington, D.C.: U.S. Forest Service.

C

Sources of Seed for Research

Table C-1 is based on responses to a questionnaire distributed by the Work Group on Managing Forest Genetic Resources during 1987–1988. It may not be comprehensive or reflect more recent changes.

TABLE C-1   Selected Agencies with Seed Available for Research Purposes

Country/Agency	Genera/Species
Argentina	
Banco Nacional de Germoplasma de Prosopis en Argentina c/o Facultad Nacional de Córdoba C.C. 509 5000 Córdoba Argentina	Arid and semiarid zone native woody species.
Instituto Forestal Nacional Pueyrredon 2446-4 Piso Buenos Aires Argentina	Some *Prosopis* species.
Australia	
Tree Seed Center Commonwealth Scientific and Industrial Research Organization (CSIRO) Division of Forest Research P.O. Box 4008 Victoria Terrace, ACT 2600 Australia	Over 900 woody Australian species, among them, *Acacia, Casuarina, Eucalyptus, Grevillea, Hakea,* and *Melaleuca*.

(continued)

TABLE C-1 (*Continued*)

Country/Agency	Genera/Species
Australia (*Continued*)	
Conservation and Land Management Department 50 Hayman Road Como W.A. 6152 Australia	Some *Acacia* species, *Casuarina cristata*, *Ceratonia siliqua*, *Eucalyptus botryoides*, *E. calophylla*, *E. camaldulensis*, and *E. globulus*.
Conservation and Land Management Department 50 Hayman Road Como W.A. 6152 Australia	Some *Acacia* species, *Casuarina cristata*, *Ceratonia siliqua*, *Eucalyptus botryoides*, *E. calophylla*, *E. camaldulensis*, and *E. globulus*.
Conservation Commission of the Northern Territory P.O. Box 38496 Winnellie, N.T. Australia	Approximately 18 genera, but generally with only one species. Among them, *Antiaris*, *Atriplex*, *Azadirachta*, *Bombax*, *Casuarina*, *Ceratonia*, *Hibiscus*, and *Tamarindus*.
Department of Conservation, Forests, and Lands G.P.O. 4018 Melbourne, Victoria 3001 Australia	Several species of *Acacia* and *Eucalyptus* and at least one species of *Acer*, *Agonis*, *Betula*, *Bombax*, *Celtis*, *Cinnamomum*, *Cotoneaster*, *Crataegus*, *Leptospermum*, and approximately 20 other genera.
Forestry Commission of New South Wales P.O. Box J 19 Coff's Harbour Jetty N.S.W. 2450 Australia	Approximately 40 genera, among them, *Acacia*, *Cinnamomum*, *Cotoneaster*, *Cryptomeria*, *Eucalyptus*, *Fraxinus*, *Gleditsia*, *Leptospermum*, *Myoporum*, *Olea*, *Paulownia*, *Photinia*, *Pyracantha*, and *Sapium*.
Queensland Department of Forestry G.P.O. Box 944 Brisbane 4001 Australia	About 35 genera, among them, *Acacia*, *Araucaria*, *Bauhinia*, *Brachychiton*, *Callitris*, *Duranta*, *Erythrina*, *Flindersia*, *Hymenosporum*, *Lagunaria*, *Melaleuca*, *Syncarpia*, and *Tipuana*.
Woods and Forests Department G.P.O. Box 1604 Adelaide 5001 Australia	Mainly indigenous species.

TABLE C-1 (*Continued*)

Country/Agency	Genera/Species

Austria

Institut fuer Forstpflanzenzuechtung
  und Genetik
Forstliche Bundesversuchsanstalt
A-1131 Wien, Schoenbrunn
Tirolergarten
Austria

All native species in Central
Europe (European Alps) and
collections from high altitudes for
Picea abies, Pinus cembra, and *Larix
decidua*. Living plant material is also
available.

Brazil

Banco de Semientos
IBDF/EMBRAPA
C.P. 1316
Brasilia, D.F. 70,000
Brazil

Eucalyptus species.

Empresa Brasileira de Pesquisa
  Agropecuária (EMBRAPA)
Centro Nacional de
  Pesquisa de Florestas
Caixa Postal 3319
80001 Curitiba-Parana
Brazil

Subtropical broadleaf native
species. *Eucalyptus* species
to develop landraces and
*Acacia mearnsii, Mimosa
scabrella.*

Florestas Rio Doce S/A
Caixa Postal 91, CEP 29000
Linhares, ES
Brazil

Pithecellobium species,
Schizolobium parahybum, and
Spondias purpurea.

Burkina Faso

Centre National de Semences
  Forestières
B.P. 2682
Ouagadougou
Burkina Faso

Data not available.

Canada

British Columbia Forest Service
Kalamalka Research Station and
  Seed Orchard
3401 Reservoir Road
Vernon, B.C. V1B 2C7
Canada

Half-sib and full-sib seed
lots of *Larix occidentalis,
Picea engelmannii, P. glauca,
Pinus contorta* var. *latifolia,*
and *Pseudotsuga menziesii.*

Canadian Forestry Service
Petawawa National Forestry Institute
Chalk River, Ontario K0J 1J0
Canada

Short- to long-term storage of
2,500 well-documented seed lots
representing 190 forest tree and
shrub species of primarily Canadian
sources; in research quantities. Seed
list available on request.

(continued)

TABLE C-1 (Continued)

Country/Agency	Genera/Species
Canada (Continued)	
Ministry of Forests and Lands Research Branch 31 Bastion Square Victoria, B.C. V8W 3E7 Canada	Pinus contorta, Pseudotsuga menziesii, and short-term storage of Picea sitchensis.
University of British Columbia Faculty of Forestry 193-2357 Main Mall Vancouver, B.C. V6T 1W5 Canada	Short-term storage of Pinus contorta and Pseudotsuga menziesii.
University of Toronto Faculty of Forestry 203 College St. Toronto, Ontario M5S 1A1 Canada	Short-term storage of some Pinus strobus, Populus deltoides, and Salix species.
Chile	
Centro de Semillas Forestales Corporacion Nacional Forestal Constitucion 291, Casilla 5, Chillan Chile	Mainly species of Acacia, Caesalpinia, Geoffroea, and Prosopis.
Instituto Forestal (INFOR) Huerfanos 554, Casilla 3085 Santiago Chile	Acacia cavenia, Prosopis chilensis, P. tamarugo, Quillaja saponaria, and other indigenous species.
China	
Chinese Academy of Forestry The Research Institute of Forestry Wan Shou Shan 10091 Beijing People's Republic of China	Mainly species of Acacia, Gmelina, Paulownia, Pinus, Populus, and Ulmus.
Colombia	
Banco Nacional de Semillas Instituto de Desarrollo de Recursos Naturales Avenida Caracas Nr. 25A66 Bogota Colombia	Mainly species of Alnus.

TABLE C-1 (*Continued*)

Country/Agency	Genera/Species

Colombia (*Continued*)

Centro Internacional de
 Agricultura Tropical (CIAT)
Apartado Aereo 6713
Cali
Colombia

Mainly species of *Leucaena.*

Instituto Nacional de Recursos
 Naturales Renovable (INDERENA)
Banco de Semillas
A.A. 13458
Bogota, D.E.
Colombia

Data not available.

Costa Rica

Latin American Forest Tree Seed Bank
CATIE
Turriabla
Costa Rica

Collections of *Alnus, Cassia,
Eucalyptus, Gliricidia, Grevillea,
Guazuma, Terminalia,* and other
Central American native species.

Cuba

Banco de Semillas
Centro de Investigaciones Forestales
Calle 174 No. 1723/17B, 17C
Siboney Marianao
Havana
Cuba

Mainly species of *Pinus.*

Cyprus

Department of Forestry
Ministry of Agriculture and
 Natural Resources
Nicosia
Cyprus

Some *Acacia* species and at
least one species of *Cassia,
Casuarina, Ceratonia, Cercis,
Hibiscus, Laurus, Parkisonia,
Pistacia, Quercus, Salix, Sophora,*
and *Tecoma.*

Denmark

Danish International Development
 Forest Seed Center Agency
 (DANIDA)
Tree Improvement Station
Krogerupvej 21
3050 Humlebaek
Denmark

Stores and distributes *Acacia* and
Prosopis for the Food and
Agriculture Organization of the
United Nations; and distributes
"semi-bulk" quantities of Central
American *Pinus* species and research
quantities of *Abies alba, A. grandis,
A. nordmanniana, Picea abies, P.
sitchensis, Pinus contorta, P. kesiya, P.
merkusii, P. sylvestris,* and *Tectona*
species. Short-term storage of
Gmelina arborea and *Tectona grandis.*

(*continued*)

TABLE C-1 (*Continued*)

Country/Agency	Genera/Species
El Salvador	
Centro de Recursos Naturales (CENREN) Apartado Postal 2265 Canton el Matasano Soyapango, San Salvador El Salvador	*Casuarina equisetifolia, Cedrela mexicana,* and *Cibistax donnell-smithii.*
Ethiopia	
Forestry Research Center P.O. Box 1034 Addis Ababa Ethiopia	Data not available.
Finland	
Finnish Forest Research Institute Department of Forest Genetics P.O. 18 SF-01301 Vantaa Finland	Short-term storage of native species of conifers and hardwoods to establish seed collection stands and seed orchards.
Finnish Forest Research Institute Kolari Research Station SF-95900 Kolari Finland	Short-term storage of native species of conifers and hardwoods to establish seed collection stands and seed orchards. Seed is mainly from north of the Arctic Circle.
France	
L'Association Forêt-Cellulose (AFOCEL) Domaine de l'etançon 77370 Nangis France	Short-term storage of *Abies grandis, Calocedrus decurrens, Eucalyptus globulus, E. gunnii, Picea abies, P. sitchensis, Pinus alternata* x *radiata, P. pinaster, P. serotina, P. taeda, Pseudotsuga menziesii,* and *Sequoiadendron giganteum.*
Centre National du Machinisme Agricole, du Genie Rural, des Eaux et des Forêts (CEMAGREF) Nogent Domaine des Barres 45290 Nogent-sur-Vernisson France	Short-term storage of some species of *Larix, Picea, Pinus, Abies alba, Cedrus atlantica,* and *Pseudotsuga menziesii.*
Centre Technique Forestier Tropical (CTFT) 45 Bis Avenue de la Belle Gabrielle F-941310 Nogent-sur-Marne France	Large collection of West African species (including *Terminalia* and dry-zone *Acacia*) and Australian species (mainly *Acacia* and *Eucalyptus*). Collection of *Pinus caribaea.*

TABLE C-1   *(Continued)*

Country/Agency	Genera/Species

France   *(Continued)*

L'Institut National de la Recherche Agronomique (INRA) INRA Bordeaux, Pierroton Laboratoire d'Amelioration des Arbres Forestiers Domaine de l'Hermitage Pierroton, 33610 Cestas France	Short-term storage of seven *Abies* species, eight *Pinus* species, and some species of *Cedrus, Cryptomeria, Cupressus, Larix, Picea,* and *Pseudotsuga.* Efforts are toward seed orchards for breeding populations and gene banks.

Germany

Institute of Forest Genetics and Forest Tree Breeding D-2070 Grosshansdorf 2 Sieker Landstrasse 2 Germany	Rangewide collection of *Picea abies* (collected from 1959–1963); initiated and supervised by Sweden; long- and short-term storage.
Niedersachsische Forstliche Versuchsanstalt Abt. C-Forstpflanzenzuchtung 3513 Staufenberg 6 Ortsteil Escherode Germany	Long-term storage of 8 conifer species and 7 hardwood species; short-term storage of 6 conifers and 16 hardwoods.
Bayer Landesanstalt fur Forstliche Saat-und Pflanzenzucht Forstamtsplatz, D-8221 Teisendorf Germany	All indigenous conifer species (e.g., *Abies alba, Larix decidua, Picea abies, Pinus sylvestris*) and several hardwoods; emphasis on high-altitude provenances.

Greece

Laboratory of Forest Genetics and Plant Breeding Department of Forestry and Natural Environment Aristotelian University of Thessaloniki Thessaloniki Greece	Short-term storage of *Abies cephalonica, Pinus brutia,* and *P. nigra.*

Guatemala

Banco de Semillas Forestales BANSEFOR/INAFO 7a Avenida 7-00, Zona 13 Ciudad de Guatemala Guatemala	Data not available.

(continued)

TABLE C-1   (*Continued*)

Country/Agency	Genera/Species

Honduras

Banco de Semillas Forestales
ESNACIFOR
Apartado 2
Siguatepeque, Comayagua
Honduras

Data not available.

Hungary

Forest Research Institute
ERTI
Experiment Station Sarvar 9600
Hungary

Data not available.

University of Sopron
Faculty of Forestry
Department of Forest Botany
H-9401 Sopron
P.O. Box 132
Hungary

Short-term storage of *Populus alba, P. canescens, P. nigra,* and *Salix alba.*

India

Forest Research Institute
P.O. Box New Forest
Dehra Dun-2048006, U.P.
India

Mainly species of *Acacia, Albizia, Alnus, Cassia, Dalbergia, Gmelina, Pinus, Prosopis, Sesbania,* and *Trema.*

National Bureau of Plant
 Genetic Resources (NBPGR)
Pusa Campus
Delhi-12
India

Species of agroforestry importance, *Acacia, Alnus, Cassia, Dalbergia,* and new *Prosopis.*

Tamil Nadu Forest Department
Forest Genetic Division
Bharathi Park Road
Coimbatore-43, Madras
India

Species of *Acacia, Alnus, Bambusa, Delonix, Eucalyptus, Hardwickia, Pongamia* (*Millettia*), *Pterocarpus,* and *Santalum.*

Indonesia

Directorate General Reforestation
 and Land Rehabilitation
Floor 13, Gedung Manggala
Wana Bakti Jin. Gatot
Subroto, Jakarta
Indonesia

Indigenous species mainly.

TABLE C-1   (*Continued*)

Country/Agency	Genera/Species

Indonesia   (*Continued*)

Perum Perhutani
Forest State Corporation
Jalan Jedral Gatut Subrota
P.O. Box 111
Jakarta
Indonesia

Species of *Calliandra* and
Leucaena.

Ireland

Forest and Wildlife Service
Department of Energy-Forest Service
1-3 Sidmonton Place
Bray, County Wicklow
Ireland

Short-term storage of some
conifer species. A few
species of *Abies, Picea,*
Pinus, and *Pseudotsuga.*

Israel

Institute for Applied Research
Ben Gurion University of the Negev
P.O. Box 1025
Beer Sheva 84110
Israel

Mainly indigenous species of *Acacia,*
Cassia, Casuarina, Cordia, Dodonaea,
Grewia, Moringa, Prosopis, Rhus,
Ricinus, Robinia, Tamarix, and
Thespesia.

Italy

Instituto Spermimentale par la
   Selvicultura
Viale S. Margherita 80
I-52100 Arezzo
Italy

Mediterranean conifer species
of *Abies, Cupressus,* and
Pinus.

Seed Unit-FAO
Plant Production and Protection
   Division
Via delle Terme di Caracalla
00100 Rome
Italy

Collection of *Prosopis*
tamarugo.

Ivory Coast

Centre Technique Forestier Tropical
(CTFT)
08 B.P. 33
Abidjan 08
Ivory Coast

Data not available.

Jamaica

Department of Forestry and
   Soil Conservation
173 Constant Spring Road
Kingston 8
Jamaica

Leucaena species (e.g., *L.*
leucocephala).

(*continued*)

TABLE C-1 (Continued)

Country/Agency	Genera/Species
Japan	
Kanto Forest Tree Breeding Institute Ministry of Agriculture, Forestry, and Fishery Kasahara-Cho Mito-Shi Ibaraki-Ken 310 Japan	Data not available.
National Seed Coordinating Centre c/o Forest Research Institute 1 Matsunosato Kukizaki-mura Inashiki-gun, Ibaraki-ken Japan	Species of *Alnus, Paulownia, Pinus,* and *Prunus.*
Kenya	
Kenya Forestry Seed Centre (KEFRI) P.O. Box 20412 Nairobi Kenya	Short- and long-term storage of *Acacia, Cupressus lusitanica, Pinus patula,* and native species of arid and tropical highlands.
Tree Seed Program Ministry of Energy and Regional Development P.O. Box 21552 Nairobi Kenya	Data not available.
Korea	
Institute of Forest Genetics Suwon Republic of Korea	Data not available.
Forest Research Institute Chungyangni-Dong Dongdaemun-Ku Seoul Republic of Korea	Data not available.
Malawi	
Forest Research Institute of Malawi P.O. Box 270 Zomba Malawi	Mainly species of *Colophospermum, Gmelina, Maesopsis,* and *Pinus.*

TABLE C-1 (Continued)

Country/Agency	Genera/Species

Malaysia (Mainland)

Forest Research Institute
Kepong, Selangor
52109 Kuala Lumpur
Malaysia

Species of *Anthocephalus,*
Cassia, and *Casuarina.*
Research quantities of
indigenous forest tree species.

Malaysia (Sabah)

Forest Research Centre
P.O. Box 1407
Sandakan, Sabah
Malaysia

Long-term storage toward
stock plantations of *Acacia*
mangium, Eucalyptus brassiana,
Gmelina arborea, and *Paraserianthes*
falcataria.

Sabah Softwoods SDN BHD
P.O. Box 137
Brumas, Tawau
Malaysia

Data not available.

Mexico

Departmento de Bosques
Centro de Genetica Forestal
Chapingo, Edo. de Mexico 56230
Mexico

Data not available.

INIFAP
Insurgents Sur 964
10 Piso, Col del Valle
03100 Mexico, D.F.
Mexico

Central American *Pinus*
species and some arid-zone
hardwood species.

Instituto Nacional de Investigaciones
   Sobre Recursos Bioticos
Apdo. Postal 63, Km. 2.5
Antigua Carretera a Coutepec
Xalapa, Veracruz
Mexico

Species of *Acacia, Erythrina,*
Inga, and *Leucaena.*

Nepal

Tree Seed Unit
Hattisar
Naxal, Kathmandu
Nepal

Data not available.

New Zealand

Forest Research Institute
Forest Health and Improvement
   Division
Private Bag
Rotorua
New Zealand

Medium-term seed storage of
Pseudotsuga menziesii, minor
collections of Mexican *Pinus* species,
and nine northern *Pinus* species.
Substantial collections of *Eucalyptus,*
Juglans nigra, and *Pinus radiata,* some
of which are unique.

(continued)

TABLE C-1 (*Continued*)

Country/Agency	Genera/Species

New Zealand (*Continued*)

National Plant Materials Centre
Aokautere Science Centre
Ministry of Works and Development
Private Bag
Palmerston North
New Zealand

Short-term storage of *Populus deltoides* collections; efforts are toward seed orchards and/or arboreta, mainly for breeding and clone banks of this species and other *Populus* species and *Salix*.

Nicaragua

Instituo Nicaraguense de Recursos
    Naturales y del Ambiente (IRENA)
Km. 12.5 Corretera Norte
Apartado No. 5123
Managua
Nicaragua

Data not available.

Nigeria

Forestry Research Institute of Nigeria
Federal Ministry of Science
    and Technology
PMB 5054
Ibadan
Nigeria

Long-term storage of *Gmelina arborea* and *Triplochiton scleroxylon*.

Pakistan

Pakistan Forest Research Institute
Forest Geneticist
Peshawar, NWFP
Pakistan

Species of *Acacia, Pinus, Prosopis,* and *Sesbania*.

Papua New Guinea

Office of Forests
Forest Research Institute
Bulolo
Papua New Guinea

Data not available.

Office of Forests
P.O. Box 5055
Boroko
Papua New Guinea

Species of *Acacia, Araucaria, Casuarina,* and *Eucalyptus*.

Peru

Banco Nacional de Semillas Forestales
Servicio Forestal y de Caza
Ministerio de Agricultura
Natalio Sanchez 20 - Jesus Maria
Lima
Peru

Caesalpinia spinosa, Casuarina cunninghamiana, Cupressus goveniana, C. macrocarpa, Jacaranda acutifolia, Poinciana (Caesalpinia) regia, Prosopis juliflora, and *Tipuana tipu*.

TABLE C-1 (*Continued*)

Country/Agency	Genera/Species

Peru (*Continued*)

Universidad Nacional Agraria
La Molina Facultad de
Ciencias Forestales
Apdo. 456, La Molina
Lima
Peru

Acacia species and *Prosopis pallida.*

Philippines

Department of Environment and
Natural Resources
Ecosystems Research and
Development Bureau
College, Laguna 4031
Philippines

Storage of small quantities of
Casuarina, Pinus kesiya, P. merkusii, Pterocarpus, rattan
species, and *Vitex.*

Puerto Rico

Southern Forest Experiment Station
Institute of Tropical Forestry
P.O. Box AQ
Rio Piedras, PR 00928
Puerto Rico

Acacia albida, Adansonia digitata, Albizia, Annona muricata, Elaeis guineensis, Flacourtia indica, Manilkara zapota, Melia azedarach, Pinus caribaea, Tectona, Terminalia, and other
genera.

Romania

Forest Research and Managment
Institute
Soseava Stefanesti 128
Bucharest 11, R-72904
Romania

Picea abies.

Rwanda

ISAR
Department de Foresterie
Centrale de Graines Forestieres
B.P. 617
Butare
Rwanda

Approximately 25 genera, among
them, *Acrocarpus, Albizia, Callandra, Callitris, Entada, Grevillea, Jacaranda, Maesopsis, Newtonia, Sesbania,* and *Tetraclinis.*

Senegal

Institut Senegalais de Recherche
Agricoles
Centre National de Recherches
Laboratoire de Graines
Parc Forestier de Hann
B.P. 2312
Dakar
Senegal

Data not available.

(*continued*)

TABLE C-1 (*Continued*)

Country/Agency	Genera/Species
Sierra Leone	
Forestry Division Manr Youyi Building, Brookfields Freetown Sierra Leone	*Leucaena leucocephala.*
Solomon Islands	
Forestry Division P.O. Box 79 Munda Solomon Islands	*Acacia margium* and *Casuarina equisetifolia.*
South Africa	
Mondi Paper Forests Division, Tree Improvement Private Bag X522, Sabie 1260 Republic of South Africa	Some *Eucalyptus* species.
South African Forestry Research Institute P.O. Box 727 Pretoria 0001 Republic of South Africa	Short-term storage of some *Acacia, Eucalyptus,* and *Pinus* species.
Sudan	
Arid Zone Forestry Research Forestry Research Institute P.O. Box 658 Khartoum Sudan	Species of *Acacia* and *Prosopis.*
Sweden	
Skogsstyrelsen National Board of Forestry S-551 83 Jonkoping Sweden	Short-term storage of *Picea abies* and *Pinus sylvestris.*
Tanzania	
Department of Forest Biology Faculty of Forestry Sokoine University of Agriculture Box 3009, Morogoro Tanzania	*Prosopis chilensis* and *Leucaena leucocephala.*
Tanzania Forestry Research Institute Silvicultural Research Center P.O. Box 95 Lushoto Tanzania	Mainly species of *Acacia, Cassia, Cupressus, Eucalyptus, Grevillea, Pinus,* and *Terminalia.*

TABLE C-1 *(Continued)*

Country/Agency	Genera/Species

Thailand

Pine Improvement Center
P.O. Hod
Chiang Mai
Thailand

Mainly *Pinus kesiya* and *P. merkusii.*

Royal Forest Department
Bangkhen, Bangkok 10900
Thailand

Mainly seeds of *Gmelina, Pinus kesiya, P. merkusii,* and *Tectona grandis.*

Teak Improvement Center
Ngao, Lampang
Thailand

Species of *Tectona.*

Tunisia

National Forest Research Institute
Route de la Souka
B.P. 2
Triana
Tunisia

Species of *Acacia, Pinus,* and *Populus.*

United Kingdom

Forestry Commission
Northern Research Station
Roslin
Midlothian, EH25 9SY
Scotland, U.K.

Short-term storage of *Pinus sylvestris.*

Oxford Forestry Institute (OFI)
South Parks Road
Oxford OX1 3RB
England

Mainly collections of Central American *Pinus* species and species of *Acacia, Albizia, Enterolobium, Gliricidia, Leucaena, Mimosa, Parkinsonia, Pithecellobium,* and *Prosopis.*

Royal Botanic Gardens
Kew (Wakehurst Place) Ardingly
Haywards Heath
West Sussex RH17 6TN
England

Some collections of forest tree species.

United States

Central America and Mexico
  Coniferous Resources
  Cooperative (CAMCORE)
School of Forest Resources
North Carolina State University
Research Annex West
Box 8007
Raligh, NC 27695-8007
USA

Short-term storage of 8 tropical hardwoods, 10 *Pinus* species, and *Abies guatemalensis*; all the material collected mainly in Central America and Mexico.

(continued)

TABLE C-1 (*Continued*)

Country/Agency	Genera/Species
United States (*Continued*)	
Department of Horticulture University of Hawaii 3190 Maile Way Honolulu, HI 96822 USA	*Leucaena* species.
National Tree Seed Laboratory 5156 Riggins Mill Road Rt. 1, Box 182-B Dry Branch, GA 31020 USA	Data not available.
Nitrogen Fixing Tree Association (NFTA) P.O. Box 680 Waimanalo, HI 96795 USA	Very complete collection of *Leucaena* species; some other nitrogen-fixing trees.
NIFTAL Project University of Hawaii P.O. Box 0 Paia, HI 96779 USA	*Afzelia africana.*
Regional Plant Introduction Station Georgia Experiment Station Experiment, GA 30212 USA	*Acacia senegal.*
U.S. Forest Tree Seed Center P.O. Box 819 Macon, GA 31298 USA	*Alnus, Fraxinus, Juglans, Paulownia, Pinus, Prunus,* and *Robinia* species.
Venezuela	
Estacion Experimental de Semillas Forestales Callejon "La Ceiba," El Limon Edo. Aragua Venezuela	*Albizia saman* and *Cedrela mexicana.*
Instituto Forestal Latino-Americano Apartado 36 Merida Venezuela	Mainly species of *Alnus* and *Pithecellobium.*

TABLE C-1 *(Continued)*

Country/Agency	Genera/Species
Zambia	
Forest Research Division P.O. Box 22099 Kitwe Zambia	Mainly indigenous species. Several species of *Acacia* and other genera, among them, *Afzelia, Berchemia, Dalbergia, Dichrostachys, Erythrina, Julbernardia, Khaya, Lonchocarpus,* and *Mimusops.*
Zimbabwe	
Forest Research Center P.O. Box HG 595 Highlands Harare Zimbabwe	Collections of species of *Eucalyptus* and *Pinus.*

Glossary

active collection Comprised of accessions that are maintained under conditions of short- or medium-term storage for the purpose of study, distribution, or use.

afforestation Establishing trees on ground where they have not previously grown.

agroforestry Any land use that combines the growing of food and tree crops. Domestic animals may be included.

allele One of two or more alternative forms of a gene, differing in DNA sequence and affecting the functioning of a single gene product (RNA and/or protein). All alleles of a series occupy the same site or locus on a pair of homologous chromosomes.

allozyme Differing forms of an enzyme, all produced by the different alleles of a single gene. They can generally be distinguished from one another using electrophoresis. (*See also* isozyme.)

angiosperm Any flowering plant; a plant bearing seeds that develop in an enclosed ovary or carpel.

apomixis Asexual reproduction in plants in which the sexual organs or related structures are involved, but fertilization does not occur. The resulting seed is vegetatively produced from an unfertilized egg or from somatic cells associated with the female parent.

base pair A pair of hydrogen-bonded chemical components (one a purine, the other a pyrimidine) that join the component strands of the DNA double helix.

biodiversity (biological diversity) The variety and variability among living organisms and the ecological complexes in which they occur.

biosphere   The largest, all-encompassing ecosystem that includes soil, water, and the atmosphere.

climax forest   The final stage in a forest succession sequence where the species composition remains relatively unchanged as long as climate and physical geography remain the same.

collecting   The general activity of gathering or acquiring genetic materials (e.g., plants, seeds) for addition to genetic resources collections. (*See also* sampling.)

community   A group of ecologically related populations of various species that occur in a particular geographic area at a particular time.

conifer   A general term referring to trees and shrubs of an order (Coniferales) that consists mostly of evergreen species; a cone-bearing tree such as pine, fir, spruce, or cypress.

cryopreservation   Maintaining tissues or seeds for the purpose of long-term storage at ultralow temperatures, typically between $-150°C$ and $-196°C$; produced by storage above or in liquid nitrogen.

deciduous   Pertains to those trees that drop their leaves at the end of the growing season, typically during winter.

dioecious   Pertains to plant species having female and male sex organs on different plants.

diploid   Possessing twice the number of chromosomes as the number present in reproductive organs such as eggs or the reproductive cells of the pollen grain. The somatic number of chromosomes (2n).

ecosystem   A community of organisms interacting with one another; the environment in which organisms live and with which they also interact.

ecotypic differentiation   The process of producing distinct individuals within a species, subspecies, or variety in a given environment that are different morphologically and physiologically from others of the same species in another environment. These so-called ecotypes cross freely with other ecotypes of the same species, subspecies, or variety and can arise from the selective pressures unique to each environment.

electrophoresis   The differential movement of charged molecules in solution through a porous medium in an electric field. The porous medium can be filter paper, cellulose, or, more frequently, a starch or polyacrylamide gel.

electrophoretic analysis   A method commonly used to separate proteins and other organic molecules.

endangered   In the context of this report, a term that applies to taxa (population, subspecies, species) in danger of extinction and for which survival is unlikely if the causal factors of loss continue.

enzyme   A large group of proteins produced by living cells that act like catalysts in essential chemical reactions in living tissues.

even-age plantation management   A practice where all trees of a plantation are of the same or very nearly the same age.

exotic   A tree growing in an area in which it does not naturally occur.

ex situ management   The management of planted stands of trees outside of their natural range; the conservation or preservation of trees as seed, pollen, tissue culture, or excised plant parts (e.g., bud cultures).

extinct   In the context of this report the term refers to taxa (e.g., populations, subspecies, species) not found after repeated searches of known and likely areas.

food chain   The pathway for energy in a natural community from producers (e.g., plants) to consumers (herbivores and carnivores) to decomposers (e.g., fungi).

food web   The complex interrelationships among the interconnecting food chains that occur within a community.

gene   The basic functional unit of inheritance responsible for the heritability of particular traits.

gene bank   An institution or center that participates in the management of genetic resources, in particular by maintaining ex situ or in situ collections; the term also can refer to a collection of genetic resources rather than the institution holding it.

gene flow   The movement of genes through or between populations as the result of outcrossing and natural selection.

gene pool   The totality of genes and their alleles within an interbreeding population.

genetic diversity   In a group such as a population or species, the possession of a variety of genetic traits that frequently result in differing expressions in different individuals.

genetic resources   In the context of this report, trees from which the genes needed by breeders and other scientists can be derived.

genotype   The genetic constitution of an individual or group that may be either expressed or unexpressed, depending on the environmental effects of a given location.

germplasm collection   A collection of many different varieties, species, or subspecies representing a diverse collection of genotypes and, hence, genetic diversity.

gymnosperm Woody vascular plants, such as conifers, that produce naked seeds not enclosed by an ovary.

habitat The place where an organism is usually found; its natural environment.

hermaphroditic Pertaining to an organism or structure that has both male and female reproductive organs.

heterozygous Having one or more unlike alleles at corresponding loci of homologous chromosomes.

homologous chromosomes Chromosomes that are morphologically similar and pair at the first division of meiosis.

homozygous Having like alleles for a particular gene at corresponding loci on homologous chromosomes.

hybridization The process of crossing individuals that possess different genetic makeups.

inbreeding The intentional or unintentional breeding or crossing of individuals that are more closely related than their parents.

in situ management The managing of organisms in their natural state or within their normal range; for trees, this may include such activities as the planting of provenance trials, seed orchards, or conservation stands within a species' natural range.

interbreeding In the context of this report, trees that are capable of actual or potential gene exchange through hybridization.

isoenzyme (isozyme) Different chemical forms of the same enzyme that can generally be distinguished from one another by electrophoresis. (*See also* allozyme.)

major genes Genes that make a large contribution to the expressed character, relative to environmental or other modifying influences (e.g., genes for flower color).

micropropagation The clonal production of trees or other plants through techniques of in vitro culture of buds, plantlets, or tissues.

minimum viable population (MVP) The size below which a population cannot remain stable or increase in number, but will decline and disappear, due either to insufficient reproduction or the genetic consequences of inbreeding.

monecious A plant species having female and male sexual organs on the same plant.

monotypic In taxonomy, having a taxon represented by one subordinate member. A monotypic genus, for example, is one in which there is only a single species (e.g., *Ginko biloba*).

morphometric Referring to the measurement or study of external form.

mutualist An organism that participates in a form of symbiosis (mutualism) in which both organisms derive benefit from the association.

Nei's measures   A measurement of genetic distance developed to estimate the number of DNA codon differences per structural gene locus for proteins and the divergence time between closely related species. It presupposes that the genes are selectively neutral (see E. B. Spiess, Genes in Populations, 2d ed. John Wiley and Sons, New York, 1989).

nitrogen-fixing bacteria   Select bacterial species that biologically convert molecular dinitrogen (N_2) to molecular forms useable by plants.

outcrossing   The breeding of unrelated plants or plants of different genotypes, usually under natural conditions.

panmictic population   A population in which mating is entirely random.

perennial crops   Crop plants that are managed to be productive over several years. They include herbaceous perennials that die back annually, such as asparagus; and woody perennials that have stems that may live for many years, such as apples, citrus crops, or mangos.

phenotype   The sum total of the environmental and genetic (hereditary) influences on a tree; the visible characteristics of a plant.

pollen vector   The carrier, such as wind or insects, of pollen from one plant or tree to another.

pollen-seed vector   The carriers or distributors of pollen or seed (e.g., wind, bird, insects).

polygenic trait   A genetically controlled characteristic that is the product of the combined interactions of numerous genes, each having individually small effects.

polymorphic   A species with several to many variable forms.

population   A group of organisms of the same species that occupy a particular geographic area or region. In general, individuals within a population interbreed with one another.

progeny testing   Determining the genetic characters or evaluating the genotype of an organism based on the performance of its offspring under controlled conditions.

provenance   Origin or source; for trees, an identifiable region in the natural habitat of a species from where the seed of the trees originally came.

provenance testing   Growing trees in different provenances to determine how they will respond. Exotics are sensitive to the use of the wrong provenance for growth, and they may die, have dieback, or produce an unacceptable form.

rare   In the context of this report, the term refers to taxa with small world populations that are not currently endangered or vulnerable, but that are at risk of loss.

reforestation   The introduction of trees on land from which they had previously been removed.

sampling   In the context of genetic resources, the use of the principles and theory of population genetics and related disciplines to ascertain what and how much collecting is needed to obtain the genetic diversity available in a population. (*See also* collecting.)

seed bank   An ex situ managed collection of seeds, usually available for distribution and exchange.

seed collection stand   Plantations or planted stands managed and maintained for the purpose of producing seed.

selection   Any natural or artificial process that permits an increase in the proportion of certain genotypes or groups of genotypes in succeeding generations in relation to others.

self-pollination (selfing)   The natural or artificial process of placing pollen grains on a receptive stigma of the same individual.

silvipastoral system   An agricultural system that combines forestry and the production of domestic animals.

species   A taxonomic subdivision of the ranking genus. A group of organisms that actually or potentially interbreed and are reproductively isolated from other such groups.

stumpage value   The value of standing timber if cut.

threatened   In the context of this report, the term refers to taxa for which the potential for loss exists but for which there is insufficient data to determine whether they are rare, vulnerable, endangered, or extinct.

tissue culture   A technique for cultivating cells, tissues, or organs of plants in a sterile, synthetic medium; includes the tissues excised from a plant and the culture of pollen or seeds.

tree   A woody perennial plant typically having a single main stem or trunk.

vulnerable   In the context of this report, the term refers to taxa believed likely to move into the endangered category in the near future if the causal factors of loss continue.

Abbreviations

ACIAR	Australian Center for International Agricultural Research
ASEAN	Association of Southeast Asian Nations, Thailand
BANSEFOR	Banco de Semillas Forestales (Forest Seed Bank), Guatemala
BCI	Barro Colorado Island, Panama
CAMCORE	Central America and Mexico Coniferous Resources Cooperative, United States
CATIE	Centro Agronómico Tropical de Investigación y Enseñanza (Tropical Agriculture Research and Training Center), Costa Rica
CGIAR	Consultative Group on International Agricultural Research, United States
CIDA	Canadian International Development Agency
CIRAD	Centre de Coopération Internationale en Recherche Agronomique pour le Développement (Center for International Cooperation on Agronomic Research for Development), France
CSIRO	Commonwealth Scientific and Industrial Research Organization, Australia
CTFT	Centre Technique Forestier Tropical (Tropical Forestry Technical Center), France
DANIDA	Danish International Development Agency
DFSC	Danish Forest Seed Center
ESNACIFOR	Escuela Nacional de Ciencia Forestal (National School of Forest Science), Honduras

FAO	Food and Agriculture Organization (of the United Nations), Italy
IBPGR	International Board for Plant Genetic Resources, Italy
ICRAF	International Council for Research in Agroforestry, Kenya
IDRC	International Development Research Center, Canada
ITTO	International Tropical Timber Organization, Japan
IUCN	International Union for the Conservation of Nature and Natural Resources, Switzerland
IUFRO	International Union of Forestry Research Organizations, Austria
MAB	Man and the Biosphere Program (of UNESCO), France
NFTA	Nitrogen Fixing Tree Association, United States
OFI	Oxford Forestry Institute, United Kingdom
TFAP	Tropical Forestry Action Plan (of FAO)
UN	United Nations, United States
UNEP	United Nations Environment Program, Kenya
UNESCO	United Nations Educational, Scientific, and Cultural Organization, France
USAID	U.S. Agency for International Development
USDA	U.S. Department of Agriculture
WCMC	World Conservation Monitoring Center (of IUCN)
WRI	World Resources Institute, United States
WWF	World Wide Fund for Nature, United States

Authors

ROBERT W. ALLARD (*Subcommittee Chairman*) Allard is emeritus professor of genetics at the University of California, Davis. He has a Ph.D. degree in genetics from the University of Wisconsin. His areas of research include plant population genetics, gene resource conservation, and plant breeding. He is a member of the National Academy of Sciences.

PAULO DE T. ALVIM Since 1963 Alvim has been the scientific director for the Comissão Executiva do Plano da Lavoura Cacaueira, Brazil. He earned a Ph.D. degree from Cornell University with specialization in plant physiology, tropical agriculture, and ecology.

AMRAM ASHRI Since 1971 Ashri has been professor of genetics and breeding at the Hebrew University of Jerusalem, Israel. He has a Ph.D. degree in genetics from the University of California, Davis. His areas of research include plant breeding and the evaluation and utilization of germplasm resources.

JOHN H. BARTON Since 1975 Barton has been a professor of law and director of the International Center on Law and Technology at Stanford University, where he earned his law degree. He is also cofounder of International Technology Management, a consulting firm specializing in international technology, trade, regulation, and transfer. He is a recognized expert on property rights as they relate to genetic resources.

KAMALJIT S. BAWA   Since 1981 Bawa has been a biology professor at the University of Massachusetts. He has a Ph.D. degree in botany from Punjab University, India. His areas of research are population biology of tropical rain forest trees, genetic variation in tropical tree populations, plant pollinator interactions, evolution of sexual systems, and conservation and management of tropical rain forest resources.

JEFFERY BURLEY   Burley has been director of the Oxford Forestry Institute, United Kingdom, since 1986. He obtained a B.A. degree in forestry from Oxford University and M.F. and Ph.D. degrees in forest genetics from Yale University. His areas of research and professional interests are tropical tree breeding, wood structure, and agroforestry.

FREDERICK H. BUTTEL   Buttel is a professor in the Department of Rural Sociology and a faculty associate in the Program on Science, Technology, and Society at Cornell University. He earned a Ph.D. degree in sociology from the University of Wisconsin-Madison. His areas of interest are in technology and social change, particularly in relation to agricultural research and biotechnology.

TE-TZU CHANG   Chang has been head of the International Rice Germplasm Center at the International Rice Research Institute since 1983 and principal scientist since 1985. He has also been a visiting professor at the University of the Philippines, Los Baños, since 1962. He earned a Ph.D. degree in plant genetics and breeding from the University of Minnesota. He had a vital role in the Green Revolution in rice. Chang has broad experience in managing and designing plant gene banks.

PETER R. DAY   (*Committee Chairman*)   Before joining Rutgers University as director of the Center for Agricultural Molecular Biology in 1987, Day was the director of the Plant Breeding Institute, Cambridge, United Kingdom. He has a Ph.D. degree from the University of London, and is a leader in the field of biotechnology and its application to agriculture.

ROBERT E. EVENSON   Since 1977 Evenson has been a professor of economics at Yale University. He has a Ph.D. degree in economics from the University of Chicago. His research interests include agricultural development policy with a special interest in the economics of agricultural research.

HENRY A. FITZHUGH   Fitzhugh is deputy director general for research at the International Livestock Center for Africa, Ethiopia. He received a Ph.D. degree in animal breeding from Texas A&M University. His field of research is the development and testing of biological and socioeconomic interventions to improve the productivity of livestock in agricultural production systems.

MAJOR M. GOODMAN Goodman is professor of crop science, statistics, genetics, and botany at North Carolina State University (NCSU) where he has been employed since 1967. He has a Ph.D. degree in genetics from NCSU, and his areas of research are plant breeding, germplasm conservation and utilization, numerical taxonomy, history and evolution of maize, and applied multivariate statistics. Goodman is a member of the National Academy of Sciences.

JAAP J. HARDON In 1985 Hardon became the director of the Center for Genetic Resources, The Netherlands. He has a Ph.D. degree in plant genetics from the University of California. His specialty is plant breeding and genetics.

VIRGIL A. JOHNSON Before his retirement in 1986 Johnson was a research agronomist for the North-Central Region of the Agricultural Research Service, U.S. Department of Agriculture, and is professor emeritus of agronomy at the University of Nebraska, Lincoln. He earned his Ph.D. degree in agronomy from the University of Nebraska. His areas of interest are wheat breeding and genetics, genetics and physiology of wheat, and protein quantity and nutritional quality.

DONALD R. MARSHALL Since 1987 Marshall has been professor of agronomy at the Waite Agricultural Research Institute, University of Adelaide, Australia. He has a Ph.D. degree in genetics from the University of California, Davis. His professional interests are population genetics, plant breeding, host-parasite interactions, and genetic resources conservation.

GENE NAMKOONG (*Work Group Chairman*) Namkoong is a pioneer research geneticist in population genetics with the Forest Service of the U.S. Department of Agriculture. He is also a professor at North Carolina State University, where he earned his Ph.D. degree in genetics. His primary research area is mathematical population genetics, particularly with respect to forest tree species.

RAJENDRA S. PARODA Paroda is the deputy director general for crop sciences at the Indian Council of Agricultural Research, New Delhi. He has a Ph.D. degree in genetics from the Indian Agricultural Research

Institute, New Delhi. He is well known for his contributions as a forage breeder and for his leadership in the field of plant genetic resources in India.

SETIJATI SASTRAPRADJA Sastrapradja is affiliated with the National Center for Research in Biotechnology at the Indonesian Institute of Science. She has a Ph.D. degree in botany from the University of Hawaii.

SUSAN SHEN Since 1989 Shen has been an ecologist for the Asia Environment and Social Affairs Division of the World Bank. She previously was a policy analyst and project director at the Office of Technology Assessment of the U.S. Congress and has extensive experience in evaluating the programs of national and international agencies with regard to forest management and conservation. She has a master of forest science degree from Yale University.

CHARLES SMITH Smith is a professor of animal breeding strategies at the University of Guelph, Canada. He has a Ph.D. degree in animal breeding from Iowa State University. His research area is in animal breeding strategies, including genetic conservation, and he has been involved in international efforts to conserve domestic animal germplasm.

JOHN A. SPENCE In 1989 Spence was appointed head of the Cocoa Research Unit at the University of the West Indies, Trinidad and Tobago. He has a Ph.D. degree from the University of Bristol, United Kingdom. His research interests are cocoa tissue culture and cryopreservation as alternatives to holding field germplasm collections.

H. GARRISON WILKES Since 1983 Wilkes has been professor of biology at the University of Massachusetts, Boston. He has a Ph.D degree in biology from Harvard University. His field of research is evolution under domestication in cultivated plants, especially maize and its wild relatives, teosinte, and the genus *Tripsacum*.

LYNDSEY A. WITHERS Since 1988 Withers has been the in-vitro conservation officer in the research program of the International Board for Plant Genetic Resources, Rome, Italy. She has a Ph.D. degree in botany from the University of Nottingham, United Kingdom, and has extensive knowledge of the application of tissue culture, cryopreservation, and plant biotechnology to the conservation of plant genetic resources.

Index

A

Abies species, 86
  A. bracteata, 60, 61, 63, 67
  A. fraseri, 63, 67, 148
  A. guatemalensis, 122, 162, 197
  in breeding or testing programs, 147–148
  endangered/threatened, 162
  genetic diversity of, 63
  seed sources for research, 188, 189, 191, 197
Acacia species, 5, 11, 47, 95
  A. albida, 119, 122, 148, 162
  A. aneura, 120, 148
  A. auriculiformis, 119
  A. mangium, 119, 120, 148, 193, 196
  A. nilotica, 95, 122, 148, 162
  A. senegal, 119, 121, 148, 162, 198
  in breeding or testing programs, 103, 119, 122, 130, 148–149
  endangered/threatened, 162
  nitrogen-fixing, 120
  seed sources for research, 183–188, 190–199
Acalypha diversifolia, 64
Acer species, 86, 149, 184
Acid rain, 34, 131; *see also* Atmospheric pollution

Acrocarpus species, 195
Adansonia digitata, 162, 195
Aesculus species, 86
Afforestation, 11, 87, 100, 201
Africa
  ex situ conservation stands, 114
  fuelwood use, 37–38
  see also specific countries
African cypress, 122
African Genetic Resources Network, 114
African Ministers Conference on the Environment, 114
Afzelia species, 162, 198, 199
Agathis species, 122, 149, 163
Agonis species, 184
Agriculture/agricultural crops
  alley farming, 42, 43
  comparing trees with, 32–33
  deforestation for, 28
  gene banks, 87
  slash-and-burn, 28
Agroforestry, 5, 115, 118
  breeding programs, 94, 96, 117, 122
  defined, 201
  research needs, 72, 96, 130
  research, training, and information dissemination, 43, 109–110
  seed collection for, 120, 190

Recent Publications of the Board on Agriculture

Policy and Resources

Managing Global Genetic Resources: The U.S. National Plant Germplasm System (1991), 174 pp., ISBN 0-309-04390-5.

Investing in Research: A Proposal to Strengthen the Agricultural, Food, and Environmental System (1989), 156 pp., ISBN 0-309-04127-9.

Alternative Agriculture (1989), 464 pp., ISBN 0-309-03987-8; ISBN 0-309-03985-1 (pbk).

Understanding Agriculture: New Directions for Education (1988), 80 pp., ISBN 0-309-03936-3.

Designing Foods: Animal Product Options in the Marketplace (1988), 394 pp., ISBN 0-309-03798-0; ISBN 0-309-03795-6 (pbk).

Agricultural Biotechnology: Strategies for National Competitiveness (1987), 224 pp., ISBN 0-309-03745-X.

Regulating Pesticides in Food: The Delaney Paradox (1987), 288 pp., ISBN 0-309-03746-8.

Pesticide Resistance: Strategies and Tactics for Management (1986), 480 pp., ISBN 0-309-03627-5.

Pesticides and Groundwater Quality: Issues and Problems in Four States (1986), 136 pp., ISBN 0-309-03676-3.

Soil Conservation: Assessing the National Resources Inventory, Volume 1 (1986), 134 pp., ISBN 0-309-03649-9.

Soil Conservation: Assessing the National Resources Inventory, Volume 2 (1986), 314 pp., ISBN 0-309-03675-5.

New Directions for Biosciences Research in Agriculture: High-Reward Opportunities (1985), 122 pp., ISBN 0-309-03542-2.

Genetic Engineering of Plants: Agricultural Research Opportunities and Policy Concerns (1984), 96 pp., ISBN 0-309-03434-5.

Nutrient Requirements of Domestic Animals Series and Related Titles

Nutrient Requirements of Horses, Fifth Revised Edition (1989), 128 pp., ISBN 0-309-03989-4; diskette included.

Nutrient Requirements of Dairy Cattle, Sixth Revised Edition, Update 1989 (1989), 168 pp., ISBN 0-309-03826-X; diskette included.

Nutrient Requirements of Swine, Ninth Revised Edition (1988), 96 pp., ISBN 0-309-03779-4.

Vitamin Tolerance of Animals (1987), 105 pp., ISBN 0-309-03728-X.

Predicting Feed Intake of Food-Producing Animals (1986), 95 pp., ISBN 0-309-03695-X.

Nutrient Requirements of Cats, Revised Edition (1986), 87 pp., ISBN 0-309-03682-8.

Nutrient Requirements of Dogs, Revised Edition (1985), 79 pp., ISBN 0-309-03496-5.

Nutrient Requirements of Sheep, Sixth Revised Edition (1985), 106 pp., ISBN 0-309-03596-1.

Nutrient Requirements of Beef Cattle, Sixth Revised Edition (1984), 90 pp., ISBN 0-309-03447-7.

Nutrient Requirements of Poultry, Eighth Revised Edition (1984), 71 pp., ISBN 0-309-03486-8.

More information, additional titles (prior to 1984), and prices are available from the National Academy Press, 2101 Constitution Avenue, NW, Washington, DC 20418, (202) 334-3313 (information only); (800) 624-6242 (orders only).